Owls of North America

OWLS
of NORTH AMERICA

FRANCES BACKHOUSE

Firefly Books

A Firefly Book

Published by Firefly Books Ltd. 2008

First printing, paperback

PUBLISHER CATALOGING-IN-PUBLICATION DATA (U.S.)
Backhouse, Frances
Owls of North America / Frances Backhouse.
[216] p. : col. photos., maps ; cm.
Includes bibliographical references and index.
ISBN-13: 978-1-77085-232-7 (pbk.)
1. Owls -- North America. 2. Owls -- North America -- Identification. I. Title.
598.9/7097 dc23 QL696.S8B334 2013

LIBRARY AND ARCHIVES CANADA CATALOGUING IN PUBLICATION
Backhouse, Frances
Owls of North America / Frances Backhouse.
Includes bibliographical references and index.
ISBN 978-1-77085-232-7
1. Owls--North America. 2. Owls--North America--
Identification. I. Title.
QL696.S8B32 2013 598.9'7097 C2013-901244-3

Published in the United States by
Firefly Books (U.S.) Inc.
P.O. Box 1338, Ellicott Station
Buffalo, New York 14205

Published in Canada by
Firefly Books Ltd.
50 Staples Avenue, Unit 1
Richmond Hill, Ontario L4B 0A7

Cover and interior design: Kathe Gray Design

Printed in China

The publisher gratefully acknowledges the financial support for our publishing program by the Government of Canada through the Canada Book Fund as administered by the Department of Canadian Heritage.

Contents

Owls and Humans

From ancient myth to Harry Potter, owls hold an enduring place in the human imagination. In some cultures they are revered, in others, feared. And for every superstition that associates owls with good fortune, a dozen more link them to mortality, sickness or evil. A small sample of the hundreds of legends, beliefs and customs that invoke owls gives a sense of the prominent and diverse roles in which these birds have been cast.

On the positive side, Aboriginal tradition in some parts of Australia holds that owls guard women's souls, and women are directed to look after their female kin by protecting owls. In South America pygmy-owls are kept as cage birds because they are believed to bring their owners luck and success in love. The Ainu people of northern Japan considered Blakiston's eagle-owl to be a divine ancestor and would drink a toast to it before setting out on hunting expeditions. Greek mythology links the goddess of wisdom, Athene, to owls, and this connection is commemorated in the name of the genus to which the burrowing owl belongs.

Associations between owls and death are prominent, widespread and sometimes very specific. In the southwestern United States, Pima Indian custom dictates that a feather molted by a living owl be placed in the hand of a dying person so that the owl can safely guide that person on the journey from life to death. In Sicily the Eurasian scops-owl is a messenger of death; its call near the house of a sick man announces that he will die within three days. For the Zapotec people of southern Mexico, the barn owl delivers the bad news and fetches the soul of the deceased. In Louisiana, Cajuns whose sleep was disturbed by the calling of eastern screech-owls used to turn their left shoe upside down or their left trouser pocket inside out to cancel this ill omen.

From a ledge on the county courthouse in Geneva, Illinois, a female great horned owl keeps watch over her nest and nestlings in a nearby tree.

The scientific nomenclature of owls reflects historical European connections between owls and sorcery. The Greek word for witch, *strix,* is used to name one genus, and its Latin derivative, *striga,* names the order Strigiformes, to which all owls belong. Owls are also associated with witchcraft in other parts of the world. Such beliefs are strong and persistent in many parts of Africa, resulting in a significant number of owl killings. Similarly, the persecution of stygian owls in Hispaniola arises from superstitions about these owls transforming themselves into witches.

The earliest known depictions of owls are found in caves in southwestern France and date back to the Upper Paleolithic period, 15,000 to 20,000 years ago. The ear tufts on an owl painted on a wall in the Crotte Chauvet cave suggest an eagle-owl or long-eared owl. In the Trois Frères cave the etched outline of a pair of snowy owls and their young recalls a time when this species occurred much farther south than it does today. A number of Australian caves also harbor ancient paintings of owls, the work of early Aboriginal artists.

Other evidence that humans have long been enthralled by these birds includes the mummified remains of barn owls in ancient Egyptian tombs. The Egyptians also used owl symbols in their hieroglyphics, as did the Mayans. Among the oldest written documents that make reference to owls are the Bible and Pliny's *Historia Naturalis,* and, somewhat later, the works of Shakespeare.

Historically, owls have not fared well at the hands of humans. Because of their alleged supernatural powers, their body parts have often been used in folk medicines and magic rituals. Some traditions, such as the widespread African custom of eating owl's eyes to improve night vision, have obvious origins, but others are more obscure. In Morocco suspicious husbands or fathers were advised to place the right eye of a Eurasian eagle-owl in the hand of a sleeping wife or daughter so that she would truthfully report on her daytime activities. Pliny, however, suggested laying the heart of a "screech owl" (the species now known as the barn owl) on the left side of a sleeping woman to induce her to reveal her own heart's secrets.

Pliny also offered a recipe to treat heavy bleeding that required boiling a barn owl in oil, then adding ewe's-milk butter and honey. In Yorkshire, England, owl soup was at one time prescribed as a remedy for whooping cough, while in Poland rheumatism was said to be cured by burning owl feathers over a charcoal fire or eating baked owl. In Uruguay burrowing owl is traditionally served to convalescents to stimulate their appetite. Chinese traditional medicine makes extensive use of owl body parts, and many owls are still killed in Asia to meet the demand.

Culinary traditions that treat owls simply as food are less common. In North America the species most commonly eaten for nourishment in the past was probably the snowy owl, which some Inuit hunters still take as game. John James

Audubon sampled the meat of a snowy owl that he had dissected for scientific purposes and declared it to be "not disagreeable eating." Great gray owls were reportedly trapped for food by some northern Native peoples.

With the colonization of North America by Europeans, owl mortality increased greatly. Most settlers had little interest in eating owls, but they didn't hesitate to kill them. As biologist Arthur Cleveland Bent noted in 1937, "Owls have few enemies except man; unfortunately they are usually shot on sight, because they are big and are picturesque as mounted specimens, or because they are supposed to destroy game and domestic poultry."

While education and legal prohibitions have largely put an end to the intentional killing of owls in North America, humans continue to exert a negative influence on many of the continent's species, with habitat destruction being the number-one cause of population declines. Some North American owls, including the great horned and the mottled, seem to be fairly tolerant of the changes humans have wrought upon the landscape over the past century, and a few species have expanded into new territory, apparently in response to habitat modifications. But even as the barred owl and the western screech-owl spread into new areas, there are hints that their numbers are dropping within their original range.

The majority of owl species in North America have a more restricted distribution and smaller populations than they did a hundred years ago. Among those whose situation is most critical are the burrowing owl, the ferruginous pygmy-owl and the spotted owl. In each case the greatest threat to the species' long-term survival is loss of vital habitat. Whether the cause is industrial logging or urban sprawl, the conversion of grasslands to croplands or the damming and diversion of rivers, the end result is the same: a place that was once a welcoming home is no longer habitable.

Ultimately, whether we can maintain the continent's full diversity of owl species and subspecies will depend on our knowledge of their particular ecological requirements and our willingness to accommodate those needs. Individuals who want to play a role in owl conservation can get involved in a number of ways. For many cavity-nesting owls, nest boxes are a satisfactory substitute for natural cavities or old woodpecker nest holes. Barn owls are especially dependent on humans to provide housing. If you have suitable habitat on your property you can put up an owl box or two at home. Or you can sign up with one of the many programs that rely on volunteers to build, erect and maintain nest boxes. Other opportunities for members of the public to contribute to owl research include reporting sightings (for example, through www.ebird.org), participating in Christmas bird counts or other surveys, and helping with banding efforts such as those that have been so central to revealing the mysteries of northern saw-whet owl migration.

Above all, you can get to know these enigmatic birds better, moving beyond myth and superstition to a deeper understanding of the fascinating realities of their lives.

CHAPTER ONE

Who's Who

Almost anywhere you go in the world you can find an owl, if you know where to look. These distinctive birds of prey populate every continent except Antarctica, and every major island. They dwell in almost all terrestrial habitats, from arid desert to arctic tundra and from dense rainforest to sweeping grasslands. Most owl species live between sea level and 8,200 feet (2,500 m), but a few inhabit high mountain terrain at elevations of up to 13,200 feet (4,000 m) in Africa and South America and 15,500 feet (4,700 m) in Tibet.

Paleontologists date the appearance of owls back to the Paleocene epoch, approximately 65 to 55 million years ago. The oldest known owl fossil is a 58-million-year-old leg bone from a long-vanished species named *Ogygoptynx wetmorei,* which was unearthed in southwestern Colorado. The modern barn-owl family emerged during the Eocene epoch, which followed the Paleocene, while other present-day owls showed up much later, in the Miocene epoch of 23 to 18 million years ago.

Along the way species came and went. Many of those that have disappeared would be barely recognizable as owls today. For example, during the Pleistocene epoch (1.5 million years ago), Cuba was home to a giant owl that stood more than 3 feet (1 m) tall and preyed on large mammals. More recently, four species of long-legged "stilt owls" resided on the Hawaiian Islands, but these became extinct some time after humans colonized the archipelago.

Although owls appear to have much in common with hawks, genetic analysis shows that their anatomical similarities are a result of lifestyle rather than shared ancestry. The owls' closest relatives belong to the order

Like this barred owl, all owls share distinctive characteristics, such as an upright posture and forward-facing eyes.

Caprimulgiformes, which includes nightjars, nighthawks, poorwills and owlet-nightjars. The latter are small Australian and New Guinean birds that bear an uncanny resemblance to owls but have small bills and slender toes.

Modern owls all belong to one order, Strigiformes, which is divided into two families: Tytonidae, which includes the barn owls and related species, and Strigidae, a diverse group collectively known as typical owls. Each family is further sorted by genus (plural: *genera*) and species.

Members of both families share a number of characteristics that make them easily recognizable as owls. These include an upright posture, a large head and compact body, soft feathers, forward-facing eyes, a small, hooked bill and powerful feet with sharp talons. Most are primarily active at night, but some are crepuscular (active at twilight) and a few are primarily diurnal (active by day).

The color of owl plumage is invariably subdued earth tones, predominantly browns and grays. No owls wear bright reds, greens or blues, and while some species have yellow or orange eyes, bills or feet, others lack even these bright accents. Because of their poor color perception, gaudy attire is of no use to owls. Instead they favor colors that make them inconspicuous during the day, when the owls of most species are at rest. As a rule, owls that live in warmer or more humid regions have darker coloring, while those that live in cooler or drier regions have lighter coloring. Subtle and intricate plumage markings also help owls blend in with their surroundings.

In most owl species males and females have identical plumage. Where there are differences, these tend to be too ambiguous for positive identification. For example, female snowy owls tend to have more barring on their white plumage than males, but there is so much variation among individuals that the most heavily barred males are indistinguishable from the whitest females.

Size can be a useful gender indicator in some owl species, especially if a male and female are seen together. In the vast majority of owl species, females are the larger sex. This pattern is known as reversed sexual size dimorphism because it is opposite to the standard avian arrangement, in which males are larger than females.

Over the years biologists have proposed more than 20 separate hypotheses to explain the prevalence of reversed sexual size dimorphism among owls (and in many daytime birds of prey), but they are far from reaching consensus on this question. Most of the hypotheses are well supported by data from some species but weakened by data from others. Furthermore, there is no agreement on how to gauge size dimorphism. Most researchers studying this question say that body mass is the most relevant measurement, but measurements of wing, tail or tarsus (lower leg bone) length and foot span have also been considered. Species may show significant dimorphism in all of these measurements, or males and females may be dimorphic in some ways and alike in others.

Males and females of most owl species have identical plumage. Snowy owls are one exception.

The barn owl's heart-shaped facial disk distinguishes it from all other North American owls.

The degree of size difference between the sexes varies widely from species to species. Of those that have been sufficiently studied to draw conclusions—which excludes many tropical species—size dimorphism is most pronounced in mottled, snowy, boreal and great gray owls. In North America the species with the smallest sexual size differences are the western screech-owl and the elf and flammulated owls. The only North American owl species in which males are the larger sex is the burrowing owl.

The most visible characteristic that separates members of the Tytonidae and Strigidae families is the shape of the facial disk. Tytonids have a heart-shaped facial disk framed by a conspicuous ruff of feathers. Strigids have a round facial disk and a ruff that either completely or partially encircles the eyes. As well, barn owls have a more elongated and compressed bill.

Close examination of the feet reveals two other features that differentiate barn owls from typical owls. On tytonid feet the second and third toes are of equal length and the claw on the third toe is serrated, with comb-like projections along the edge. Strigids have a longer third toe and all of their claws have smooth cutting edges.

The two families are also distinguished by several skeletal differences. In barn owls the "wishbone" (technically known as the furcula) is fused to the breastbone, or sternum, which bears two notches. In typical owls the furcula is separate and there are four notches in the sternum. Typical owls also have a rounder skull with larger eye sockets than barn owls.

Family Tytonidae: Barn Owls and Relatives

The small Tytonidae family has only two subgroups. The genus *Tyto,* with 14 species, includes barn owls, grass owls, sooty owls and masked owls. The genus *Phodilus* consists of two rare species of bay owls, one found in southeast Asia and the other in Africa.

Tytonids live mainly in tropical and subtropical areas. They are found as far south as southern South America and Australia, but no species in this family inhabit northern regions of North America or Eurasia. The only tytonid that lives in North America is the barn owl, which is sometimes called the common barn owl to differentiate it from its kin. This species is the world's most widely distributed and most intensively studied owl.

Like most other members of the genus *Tyto,* the barn owl has a large, rounded head and a tapered body with a short, squared tail, which gives it a top-heavy appearance when flying. Its long wings and legs are noticeable in flight. The plumage of *Tyto* owls ranges from black or dark brown through light gray or buff to pure white. It is typically darker on the upper surfaces, which are delicately patterned with elaborate markings, and lighter underneath. The underparts may be unmarked, spotted or barred, but are never streaked.

Family Strigidae: Typical Owls

Worldwide, the Strigidae family includes 189 living species within 25 genera, as well as 4 extinct species. Within North America there are 22 species in 11 genera. Taxonomists frequently revise the classification of strigid owls as they gain new insights into genetic relationships. Genus names are changed, species are regrouped and subspecies are elevated to full species status or vice versa. In 2002, for instance, the genus *Nyctea,* whose only member was the snowy owl, was merged with the genus *Bubo,* and the snowy owl's scientific name was changed from *Nyctea scandiaca* to *Bubo scandiacus.* In the same year the screech-owls were separated from the genus *Otus* and assigned to the newly created genus *Megascops*. In 1997 three subspecies of the least pygmy-owl were recognized as full species, among them the Colima pygmy-owl of northern Mexico.

Because members of each genus are closely related, they generally have physical similarities to each other and may also share behavioral traits. In some cases familiarity with generic characteristics can be a useful aid to species identification. All typical owls have plumage in shades of brown, gray, buff and cream, as well as black and white. Markings include spots, bars and streaks, often in combination.

The genus *Otus* has only one North American representative, the flammulated owl. In common with its *Otus* relatives, which are collectively known as scops-owls, the flammulated owl has conspicuous ear tufts, short legs and a brief, simple song. Members of this genus are small, like the flammulated, to medium-sized.

Megascops is an exclusively New World genus, represented by four species in North America: the eastern, western, vermiculated and whiskered screech-owls. These small owls have prominent ear tufts, short legs and plumage patterns that are remarkably consistent among all four species. Their long, complex songs separate them from the *Otus* owls. Until 1983 the western and eastern screech-owls were considered to belong to a single species. They were separated on the basis of subtle plumage differences and significant disparities in their vocalizations.

The genus *Bubo* includes the world's largest owl, the Eurasian eagle-owl, which can weigh 8.8 pounds (4 kg) or more. The two North American representatives of this genus, the great horned and snowy owls, are also heavyweights. The record for a great horned owl is 5.5 pounds (2.5 kg), but weights of 2.8 to 3.7 pounds (1.3–1.7 kg) are more typical. Average weights for snowy owls are about 4 to 4.8 pounds (1.8–2.2 kg). Besides being big and powerful, *Bubo* owls also share the trait of having completely feathered legs and feet.

Members of the genus *Strix* are relatively large owls that live in forested habitats. In North America this genus is represented by three species: the great gray, barred and spotted owls. The great gray owl is North America's largest owl in terms of body length and wingspan (although large snowy owls surpass small great grays in these dimensions), but this species ranks third in terms of body mass. Great gray owls weigh at least 15 percent less than great horned owls; they look bulkier because of their long, fluffy plumage and disproportionately large facial disk.

With its subtly patterned plumage, this western screech-owl might easily be mistaken for a branch stub on a tree.

Ciccaba owls resemble members of the previous group in many ways, and some taxonomists say they should be part of the genus *Strix*. However, the

False eyespots create a surprisingly realistic impression that this northern pygmy-owl has eyes in the back of its head.

American Ornithologists' Union currently considers them distinct. The genus *Ciccaba* is restricted to the New World and has only one North American representative, the mottled owl.

The genus *Surnia* consists of only a single species, the northern hawk owl, which lives in both North America and Eurasia. This long-tailed, medium-sized owl is one of the world's least nocturnal owls.

Another genus with only one member is *Micrathene*. Its sole species is the elf owl, a resident of desert and open woodland habitats in the southwestern United States and northern Mexico. Weighing only 1.2 to 1.9 ounce (35–55 g)—about the weight of a golf ball—this is the world's smallest owl.

Members of the genus *Glaucidium* are small owls that are most active around dusk and dawn or during the day. Classification of species and subspecies within this genus has been subject to many revisions, with more likely to come. Currently taxonomists count three species of *Glaucidium* owls in North America: the Colima, ferruginous and northern pygmy-owls. A distinguishing characteristic of pygmy-owls is their "false eyes"—two white-rimmed oval black spots on the back of the neck. As well, pygmy-owls have relatively long tails, which they usually hold straight down or angled slightly out from the body. When agitated they jerk their tails up and down and from side to side.

The genus *Athene* has one representative in North America, the burrowing owl. In the past this species was assigned to its own genus, *Speotyto,* because it differs from other *Athene* owls in having longer legs and nesting in underground burrows. In 1997 the American Ornithologists' Union decided there were enough similarities in bone structure, general appearance and genetics for the burrowing owl to be placed in its present genus.

Members of the genus *Aegolius* are small to medium-sized owls with short tails, no ear tufts and somewhat square-looking heads. Anatomically they are distinguished by significant asymmetry in the size and shape of their right and left ear openings. The two *Aegolius* species that live in North America are the boreal and northern saw-whet owls.

Members of the genus *Asio* are medium-sized owls that can be divided into two groups: those that have long ear tufts and live in woodlands and those that have short ear tufts and inhabit grasslands and other open habitats. The names of two of North America's three *Asio* owls—the long-eared and short-eared—clearly indicate which group they belong to. The third in this group, the stygian owl, is of the long-ear-tuft type.

CHAPTER TWO

Built for the Night Shift

From head to toe, owls are superbly designed for their lives as birds of prey. Like hawks and eagles they are equipped with highly effective tools for killing and dismembering their prey: strong feet with curved, stiletto-like talons and a sturdy hooked bill with razor-sharp cutting edges. But what makes owls unique is that most of them hunt at night or during the twilight hours around dusk and dawn. The anatomical adaptations that allow them to work the night shift include keen hearing, enhanced low-light vision and sound-muffling structures on their flight feathers.

Most owls also establish and defend breeding territories and nest sites, find mates, form pair bonds and care for offspring during the hours when a majority of other birds are sleeping. Sensory specialization is certainly a key factor in their ability to move with ease in a darkened world, but spatial memory also plays a critical role. That's why the most nocturnal owl species—especially the ones that hunt and nest in spatially complex habitats such as forests—are also the most territorial and sedentary. Because they stay in one small area all their lives and become intimately familiar with every part of it, these owls can use their powers of recall to navigate by the merest glimmer of starlight or when even less illumination is available.

In contrast, owl species that are more active by day or in the evening and early morning tend to have one or more of the following characteristics: some degree of flexibility in territorial attachments; migratory or nomadic behavior; and a preference for open, spatially simple habitats, at least for hunting. These owls may use spatial memory to some extent to help avoid obstacles and find their way after dark, but their mental maps are drawn over the course of a few days or months rather than a lifetime.

As the sun sets, a great horned owl prepares for a night of action.

The great horned owl's eyeballs are as large as an adult human's.

Vision

Owls cannot see in total darkness—no animal can—but they can see better under low-light conditions than birds that confine their activities to daytime. One reason for owls' superior night vision is the size of their eyes. All birds have large eyes relative to their body size, but owls have maximized the size of these sense organs. A great horned owl's eyeballs are as large as an adult human's.

Large eyes allow room for a large, globular lens, which produces a bigger retinal image than the more flattened, disk-like lens found in the eyes of most other birds and humans. More important, in a larger eye the size of the pupil (the opening that admits light) can be bigger; as the amount of light entering the eye increases, so do the brightness, size and sharpness of images projected onto the retina, the light-sensitive membrane that lines the back of the eye.

Another important difference between the eyes of owls and those of day-time birds is the design of the retina itself. A retina contains two types of light receptors: cones, which work well in bright light and are responsible for color perception, and rods, which work well in dim light and discern mainly shades of gray and black. While most birds, as well as humans, have more cones than rods in their retinas, the reverse is true of owls and other nocturnal birds. As a result, owls are somewhat limited in their ability to see colors—they may confuse red with dark gray, for example—but they gain an advantage in perceiving form and detail in low-light conditions.

A number of biologists have tried to determine the minimum amount of light that owls need for hunting by running experiments in which owls attempted to find dead mice under varying levels of illumination. By using an experiment room with a sand-covered floor, the researchers could tell when the test subject flew straight from its perch to the mouse—indicating that it could see its prey from a distance of at least 6 feet (1.8 m)—and when it landed and either searched up close or wandered randomly until it found (or didn't find) the mouse. Although the researchers could not precisely quantify the owls' visual thresholds, the experiments did provide a rough measure of the minimum amount of light the subjects required to see stationary objects. Moving prey would be more readily detectable.

In two separate studies of this type, Lee R. Dice and Carl D. Marti found that the lower limit of vision in dim light for barred, long-eared, barn and great horned owls was around 0.00000073 foot-candles, equivalent to the illumination produced by a single candle at a distance of 1,170 feet (357 m). This means that these owls could readily spot prey on the floor of a deciduous forest on a clear, moonless summer night. However, on a cloudy and moonless summer night in the same forest, they would have to be within a few feet of the mouse to see it. Burrowing owls tested by the two researchers could not see in as low a light range as the other four owls, being unable to find the mice directly at light intensities of less than 0.000026 foot-candles. Although burrowing owls hunt only in open country, a moonless night with heavy cloud cover would reduce illumination levels below this threshold in such habitats.

Biologist Graham Martin, an expert in avian sensory science, used a different approach to assessing owls' visual sensitivity. He trained tawny owls to sit on a perch in front of two translucent panels that could be darkened or lit to varying intensities, and to indicate which panel was lit by pecking at a metal bar. When the illuminated panel was so dim that the owl did not respond, it was considered to have reached its visual threshold.

Martin found that the tawny owl's visual threshold is about a hundred times lower than a pigeon's—a typical daytime bird—but about twice as high as a domestic cat's, even though both owls and cats are nocturnal. The cat's superior nighttime vision is due largely to a structure within the eye called the tapetum lucidum, which reflects light back to the retina. Cats and most other nocturnal mammals have a tapetum, which is why their eyes glow when caught in a car's headlights, but the only birds with this feature are those in the nightjar family. Nevertheless, some owls show amber to reddish "eye-shine" when illuminated at night. This is caused by light reflecting off blood vessels on the retina's surface, the same as the "red-eye" effect seen in flash photographs of people.

Martin also determined that the tawny owl's visual threshold is about twice as low as the average human's; in other words, owls have about double the visual sensitivity at low light levels. That may seem like a surprisingly small difference, but Martin points out that we humans rarely give our own night vision a fair test (it takes at least 40 minutes for our eyes to fully adjust to darkness) or spend

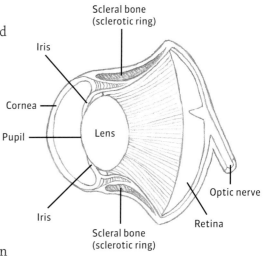

Cross-section of an owl's eye.

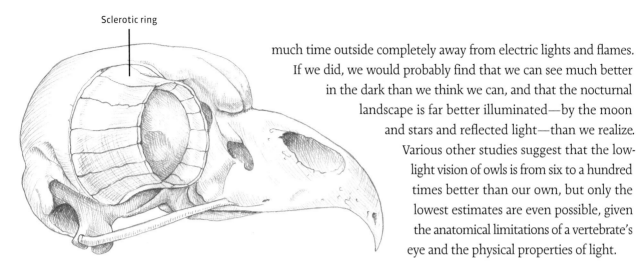

The sclerotic ring.

much time outside completely away from electric lights and flames. If we did, we would probably find that we can see much better in the dark than we think we can, and that the nocturnal landscape is far better illuminated—by the moon and stars and reflected light—than we realize. Various other studies suggest that the low-light vision of owls is from six to a hundred times better than our own, but only the lowest estimates are even possible, given the anatomical limitations of a vertebrate's eye and the physical properties of light.

Besides being designed to give a visual edge at night, owls' eyes are also made for spotting and zeroing in on prey, regardless of light levels. When we look at an owl's eyes we see only the iris and the pupil. This viewpoint offers no hint of the eye's true form, which is tubular rather than ball-shaped. Tubular eyes can accommodate the owl's large, spherical lens and are good for long-distance vision. Northern hawk owls and great gray owls, for example, can make out a mouse scurrying across the snow from half a mile (800 m) away.

On the other hand, owls, especially large ones, are somewhat limited in their ability to focus close up. Optical testing of 15 species of owls by C.J. Murphy and H.C. Howland revealed that the nearest focal distance for snowy owls was 5.5 feet (166 cm), while for great gray owls it was 2.8 feet (85 cm) and for great horned owls 1.7 feet (50 cm). Most of the small to medium-sized owls they tested had near-point focus distances of about 10 inches (25 cm), with the notable exception of the barn owl, which could focus on objects as close as 4 inches (10 cm) away. The majority of the 15 species could focus at a distance of less than 3.3 feet (1 m) for a brief period, but the only two that could maintain their near-point focus were the barn owl and the collared scops-owl. The others typically focused on an object while it was in motion and then resumed their usual distant-vision focus once the object stopped moving. Because of their tubular shape, owls' eyes extend well out from the skull. Each eye is surrounded by a circle of plate-like bones, called a sclerotic ring, that supports the eye but also makes it virtually immobile. Unlike us, owls cannot roll their eyes from side to side or up and down to increase their field of vision. They compensate for this restriction with a range of head movement that is the envy of many stiff-necked humans. With 14 cervical vertebrae (double the number found in the human neck), an owl can rotate its head more than 180° from center front to look backward over either shoulder. It can also turn its head completely upside down to get a sharper view of overhead objects.

A sideways perspective helps this short-eared owl get a sharper view.

WARNING

BURIED FIBER
OPTIC CABLE
IN THIS
VICINITY

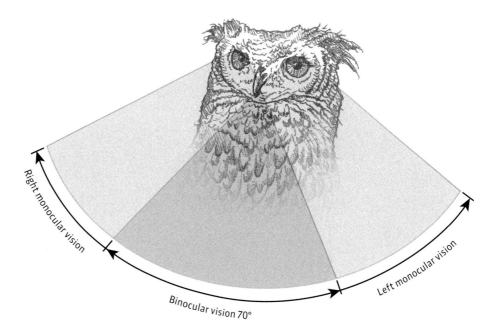

Right monocular vision

Left monocular vision

Binocular vision 70°

Binocular vision.

Owls have relatively flat faces, with their eyes positioned in the front of the head, looking almost straight forward. This placement gives them a more limited field of view than birds whose eyes are set on the sides of the head, but enables a wide area of overlap between the visual field of each eye and, therefore, good binocular vision—the ability to see three-dimensionally. Binocular vision is a great asset to owls because it gives them excellent depth perception, allowing them to pinpoint the exact position of their victims as they swoop in for the kill.

Because owls have binocular vision only across the central 50 to 70 percent of their field of view, they sometimes make head movements to help them determine an object's position, viewing it from slightly different angles and slightly different locations relative to its surroundings. This process, called parallactic localization, is essential to birds that have poor binocular vision—such as pigeons, which constantly jerk their heads back and forth as they walk. Head actions typically used by perched owls to augment their binocular vision include vertical bobbing, lateral movements and weaving.

Part of the mystique of owls may arise from the placement of their eyes, which lends them a somewhat human appearance. This resemblance is heightened by the fact that owls usually lower their upper eyelids when closing their eyes to sleep, unlike most other birds, which raise their lower lids. Owls look particularly human when they drop one eyelid in a wink.

Owls use their regular eyelids to close their eyes when sleeping or roosting and when engaged in certain activities that may put their eyes at risk, such as scratching, preening, copulating and transferring food from bill to bill between mates or parents and offspring. For blinking, however, owls (like other birds) have a third lid on each eye, called the nictitating membrane. Frequent flicking of this thin membrane across the eye keeps the cornea clean and moist. Owls also use the nictitating membrane to protect their eyes when striking prey.

A western screech-owl lowers its upper eyelids as it drops off to sleep. Most birds other than owls close their eyes by raising the lower lids.

Hearing

The feathered ear tufts that characterize a number of owl species have nothing to do with hearing; they are discussed in detail later in this chapter. Other head plumage, however, does play an important role in how owls hear. All owls have a feathered facial ruff and facial disk. Together these features give barn owls their distinctive heart-shaped face and typical owls a more rounded face. Owls that rely heavily on hearing to locate prey obscured by darkness or hidden by snow, leaves or grass, such as barn, great gray and short-eared owls, have very well-defined ruffs and disks. In species that hunt more by sight, including the snowy owl and pygmy-owls, the ruff and disk are less distinct and may be incomplete in the area of the forehead, chin or both.

The facial ruff is made up of layers of stiff, dense, extremely closely packed feathers that curve around behind the ears to form a pair of parabolic reflectors, similar to TV satellite dishes, on either side of the head. These relatively solid concave surfaces collect and focus sound waves, acting as amplifiers. The barn owl's ruff, for example, has a tenfold sound amplification effect.

The facial disk, situated in front of the ear openings, is formed by sparse feathers with very open vanes that radiate out from each eye and overlap the ruff. These "sound-transparent" feathers allow sound waves to pass through easily while preventing small objects and debris from entering the ear canal. They probably also reduce flight noise caused by air turbulence around the ear opening. The feathers of the facial disk usually do not fully cover the ruff, so its outer edge remains visible, often as a dark border around the face.

Behind the feathers of the facial disk are the ear openings. Birds typically have tiny, round ear holes, but in most owls the open end of the ear canal is very large and usually oval or roughly crescent-shaped. Compared to other birds, owls also have larger internal ear parts and a higher concentration of auditory neurons in the brain. The ear opening in the skin that covers an owl's head does not always match the ear opening in the skull—in either shape or size. In barn owls the ear openings in the skin are small and nearly square, whereas *Asio* and *Aegolius* owls have skin openings that are long slits, extending from below the lower jaw to the top of the head.

Around each ear opening most owls have only a narrow fold of skin. The feathers at the front of the facial ruff are attached to this ridge of skin and the rest of the ruff feathers are packed behind it. Some species have movable ear flaps that help direct sounds into the ear. These fleshy folds of skin may be located in front of the ear openings and may overlap the openings to a greater or lesser extent. Or they may be situated behind the openings and form a broad, forward-facing rim to which the feathers of the front edge of the ruff are attached. *Asio* owls are among the species that have ear flaps in both positions.

Owls and humans can hear sounds across a similar range of frequencies. At certain frequencies, however, owls have much more acute auditory perception than we do and are able to detect sounds that are inaudible to humans. The range of greatest aural sensitivity for owls varies between species, from about 0.5 kHz at the low end to 6 to 9 kHz at the high end. At these frequencies they can hear up to 10 times better than humans.

Along with their keen hearing, owls also have the ability to pinpoint exactly where a sound is coming from. The species that do this best are aided by a unique adaptation—asymmetrical ears. Most owls, like humans, have symmetrical ears (that is, the right and left ear are mirror images of each other) and are fairly adept at localizing sounds. In a process called binaural fusion, our brains compare minute differences in the arrival time and intensity of sound waves

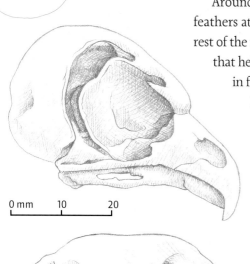

0 mm   10   20

Front and side view of boreal owl (Aegolius funereus) skull and ears.

as they reach each ear and translate the differences into spatial information about the source of the sound.

If a sound is coming from the listener's left, it will reach the left ear slightly sooner than the right and with slightly greater intensity. By turning the head until the sound waves reach both ears at exactly the same moment and with equal intensity, the listener ends up facing the source of the sound. Owls, with their large heads (necessary to accommodate their large ears and eyes), are well designed for binaural fusion, as this process works most effectively when the ears are positioned far apart.

An animal with symmetrical ears can localize sounds in the horizontal plane and determine if the source is to the left or the right. However, it must tilt its head to determine vertical direction—whether the sound is coming from above or below. To get a precise location the brain must store and combine the two sets of information, an act that cannot be carried out quickly enough to keep track of a fast-moving sound source such as a running mouse. Asymmetrical ears enable owls to apply the binaural process in both the vertical and horizontal planes simultaneously and rapidly, without moving the head. This saves time and eliminates directional uncertainty about moving prey. Because of anatomical limitations, owls can determine vertical direction for high-frequency sounds only. Fortunately for these hunters (but not for their victims), the rustling noises made by small animals moving in vegetation or snow contain both high- and low-frequency components.

Worldwide, at least 42 owl species in eight genera have evolved some form of ear asymmetry, giving them a significant advantage in sound localization. There are five basic patterns of ear asymmetry, all of them represented among the North American owls.

In barn owls and other members of the genus *Tyto,* the ear openings in the skull are symmetrical in size, shape and position, but the openings in the skin are at different levels, with the one on the left higher than the one on the right. There is also a difference between the two ear flaps in front of the openings. The left flap is higher and angled downward, so the left ear is more sensitive to sounds coming from below the head, while the right flap is lower and angled upward, making the right ear more sensitive to sounds coming from above.

Members of the genus *Aegolius*—the northern saw-whet and boreal owls in North America—have a completely different approach to asymmetry. Their ear

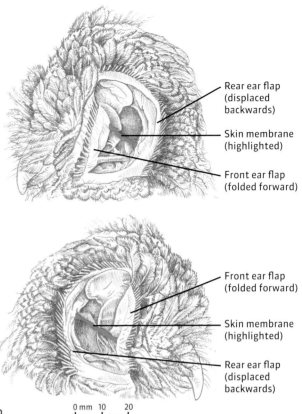

Rear ear flap (displaced backwards)

Skin membrane (highlighted)

Front ear flap (folded forward)

Front ear flap (folded forward)

Skin membrane (highlighted)

Rear ear flap (displaced backwards)

0 mm 10 20

Left and right side views of long-eared owl (Asio otus) ears.

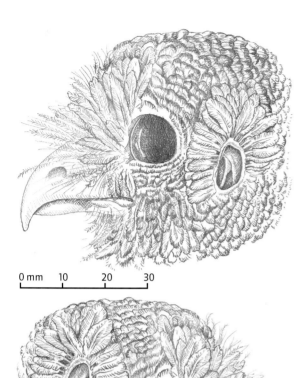

0 mm 10   20   30

Left and right side views of mottled owl (Ciccaba virgata) ears.

flaps and the long, slit-like openings in the skin are perfectly symmetrical, but the bony parts of the outer ear are conspicuously skewed. The right opening in the skull is higher than the left, and the right ear canal is directed upward while the left ear canal is horizontal.

Strix owls combine modifications of both the fleshy and bony parts of the ear. The most complex version of this form of asymmetry is the great gray owl's. The ear openings in the great gray's skull are of different shapes and the left ear canal is directed more upward than the right. As well, the ear openings in the skin are of different sizes (the right one is larger) and the flaps in front of them are of different shapes. Finally, a horizontal skin membrane cuts across each opening, but with slightly different placement on each side, influencing the way sound waves enter the ear.

In *Asio* owls, including the long-eared, short-eared and stygian, the fleshy and bony parts of the ear are entirely symmetrical, except for one modification: a skin membrane stretches nearly horizontally across the opening of each ear canal, making the left opening higher and upwardly directed, and the right opening lower and aimed horizontally.

The mottled owl and other members of the genus *Ciccaba* lack ear flaps and have very simple ear asymmetry, with the right ear opening in the skin being about one and a half times longer than the left opening.

The anatomical details of owl hearing can be difficult to grasp, but the results they produce are easy to understand. Experiments have shown that barn owls, northern saw-whet owls and possibly long-eared owls are capable of catching mice in the complete darkness of a light-tight room, even though owls never encounter total darkness in the real world. Another, more natural demonstration of the auditory prowess of owls is given by species that use only their hearing to detect rodents moving underneath a cover of snow or vegetation—as much as 18 inches (45 cm) below the surface in the case of great gray owls.

Sense of Smell

One of the unanswered questions about owls' sensory capacities is whether they use olfaction, or the sense of smell, to detect prey. One study of avian olfactory bulbs found that this part of the brain is relatively well developed in

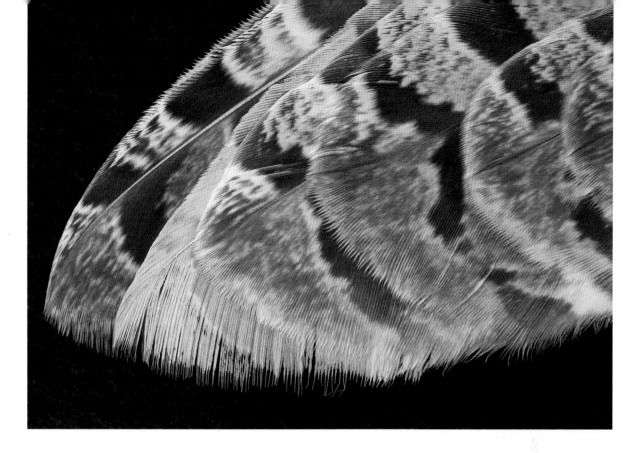

short-eared owls. A short-eared owl's olfactory bulb is about the same size as a pigeon's or a domestic chicken's and is more developed than a starling's. Since the three latter species are known to be capable of discriminating between subtly different scents, the same may be true of owls.

Owls are unable to locate dead prey in the absolute darkness of a light-tight room, which suggests that their sense of smell is not strong enough to hunt by. However, they might use olfactory cues for other purposes, such as to detect certain prey at close range or to accept or reject prey once it is caught. Support for this idea comes from an experience reported by ethnologist and naturalist Lucien McShan Turner after an expedition to Labrador in the 1880s. In the course of his research Turner collected a number of small birds, which he wrapped in paper and hung up in a cloth bag. Although the birds were completely concealed, a great horned owl found the bag, tore it open and ate more than a dozen of the specimens.

Silent Flight

Because owls aren't the only nocturnal animals with keen hearing, they have evolved the ability to fly virtually silently so they can sneak up on their prey without being heard. Silent flight further benefits owls by damping aerodynamic noise that would interfere with their efforts to hunt by ear. Not only is owl flight inaudible to humans and other owls, it is also silent in the ultrasonic range, an essential adaptation, since small mammals can detect the very high-frequency sounds of this range.

Owl flight feathers have three structural adaptations that contribute to this stealthy flying. First, every feather that forms part of the wing's leading edge has a stiff, comb-like edging along its front margin. Second, along the wing's trailing edge is a soft, hair-like fringe created by very supple feathers with non-interlocking barbs (the individual filaments that make up the vane). These two features mute aerodynamic noise by altering the way air flows over and under the wing. The third feature—a velvety layer that covers the upper surface of the flight feathers—diminishes sounds made by the feathers sliding past one another as the wing extends and flexes.

These silencing structures are not equally well developed in all species. Pygmy-owls and elf and flammulated owls all fly with quiet but perceptible wing-beats, though for different reasons. Pygmy-owls hunt mostly in the daytime or twilight hours, when silence is less advantageous. Elf and flammulated owls are strictly nocturnal but feed almost exclusively on insects, which pay little heed to the sounds of flying predators.

Ear Tufts

Approximately a quarter of the world's owl species have two tufts of feathers on the crown of the head that are longer than the surrounding feathers and resemble ears or horns. These features are commonly referred to as ear tufts. In North America, species with conspicuous ear tufts include the aptly named great horned and long-eared owls, the stygian owl and all of the screech-owls. Short-eared and flammulated owls have small ear tufts that remain concealed most of the time. Northern pygmy-owls lack typical ear tufts but can create the appearance of tufts even though their crown feathers are all the same length. They do this by erecting some head feathers while compressing the rest.

When visible, ear tufts are a helpful distinguishing characteristic for humans trying to identify owls, but biologists have yet to determine whether they are used for species recognition by the owls themselves. Some of the species with prominent ear tufts are strictly nocturnal forest-dwellers. In the dark environments where they live, vocal cues to identification are much more useful than visual ones. Other ear-tufted owls spend some of their time in fairly open habitats, where there is more ambient light. For these owls, ear tufts possibly play a role in species recognition.

Another proposition about the function of ear tufts is that they help owls repel mammal predators from their nests by mimicking the predator's ears and making

Prominent ear tufts give the
long-eared owl its common
name.

the defending owl look like a more dangerous adversary. The fact that seven species of ear-tufted owls are found only on remote islands where there are no mammals weakens this hypothesis, but it has not been either proved or disproved.

Some biologists believe that ear tufts help camouflage tree-roosting owls by breaking up the round outline of the head and making it blend in better with adjacent branches and bark. Supporting evidence for this idea includes the tendency of many ear-tufted species to assume an upright posture with tufts fully erected when they are disturbed by a potential enemy while roosting in trees or shrubs. However, not all observers agree that this so-called concealing posture actually makes owls less noticeable (see chapter 7 for more details).

Whatever other functions they may serve, ear tufts seem to play some role in visual communication, since owls often erect or flatten their tufts during encounters with mates, competitors and predators.

CHAPTER THREE

Feeding Habits

When we imagine an owl hunting, most of us probably picture it swooping down in the dark with talons outstretched toward a doomed mouse that may never know what hit it. This is exactly how many owl meals are obtained, but there is much more to owl feeding habits and behavior than this single scenario. Hunting strategies vary from species to species, and freshly caught meat is not the only item on the menu.

Prey

Owls are almost exclusively carnivorous. The only known exception is the little owl, a Eurasian and north African species that occasionally eats grasses, foliage and small fruits. However, not eating plants still leaves owls with plenty of dining options. They eat every class of vertebrate—mammals, birds, reptiles, amphibians and fish—and numerous kinds of invertebrates. Except when females and young are fed by males, owls nearly always kill their own prey. Among the species that occasionally feed on carrion are the northern pygmy-owl, the northern saw-whet owl and the great horned, northern hawk and burrowing owls.

In North America the mammals most commonly killed by owls are mice, rats, voles and lemmings. However, any terrestrial mammal up to the size of a large hare or an opossum is a possible target, and several owls, particularly the stygian, catch bats in flight. The largest mammal prey are those killed by great horned owls. With their powerful talons, which take

With lethal precision, a northern saw-whet owl zeros in on an unlucky deer mouse.

29 pounds (13 kg) of force to open, these owls can break the spines of animals even larger than themselves.

The largest birds taken by great horned owls are ducks, geese and herons. Snowy owls also kill large waterfowl, up to and including medium-sized geese. Because most birds are not active in the dark, those that fall prey to highly nocturnal owls generally belong to one of three categories: other nocturnal birds (including smaller owls and certain burrow-nesting seabirds), species that roost in the open or sit on open nests at night, and nighttime migrants. Less nocturnal owls take a wider variety of avian prey. Owls of different species catch birds while they are variously perched, flying, on the ground or on water. Pygmy-owls sometimes extract the young of woodpeckers, wrens and other cavity-nesters from their nests.

Many North American owls occasionally eat reptiles, but only diurnal owls that live in relatively warm areas eat significant numbers of these prey. The reptiles most commonly hunted by owls are lizards and their relatives. In southern Texas small lizards comprise about one-fifth of the diet of the primarily insectivorous ferruginous pygmy-owl, a greater proportion than either birds or mammals make up. Other reptilian prey include baby alligators (killed by barred owls in the southeastern United States), turtles (irregularly taken by barred and burrowing owls) and snakes. The latter can be dangerous adversaries, even for large owls. More than one great horned owl has caught a snake and then become immobilized when its intended victim coiled itself around the bird's body and neck, preventing it from either flying away or using its bill to deliver a killing bite. The resulting impasse sometimes ends fatally for both participants.

Few owls eat amphibians and most of them do so only rarely. Adult frogs and toads are the most common amphibian prey. Eastern screech-owls also eat tadpoles and, along with barred and burrowing owls, they are among the few owls known to eat salamanders.

The only specialized fishers among the owls are members of the genera *Ketupa* (Asian fish-owls) and *Scotopelia* (African fishing-owls). However, a few other species have developed a knack for catching fish and other aquatic prey such as crayfish and leeches. In North America these include the barred, great horned and snowy owls and the screech-owls.

Because insects and other invertebrates are scarce in the winter in northern regions, dependence on these prey tends to be highest among owls that inhabit more southerly latitudes or that migrate to warmer areas. Some northern species feed heavily on invertebrates in the summer and switch to vertebrate prey in winter. In North America the most insectivorous owls are the flammulated, elf, burrowing, mottled and ferruginous owls and the vermiculated and whiskered screech-owls. Insects commonly eaten by owls include grasshoppers, crickets, moths, beetles and cockroaches.

North American owls also consume a wide variety of non-insect invertebrates,
including earthworms, snails, centipedes and scorpions, but rarely in great quantities. Northern saw-whet owls living on the Queen Charlotte Islands of northern
British Columbia are an exception. Sand-hoppers (amphipods) and other beach-
dwelling invertebrates are an important source of nourishment for these owls
during the non-breeding season and may constitute as much as 15 percent of
their total annual food intake, the rest being mainly small mammals. One adult
female from this population that was killed by a car had 156 sand-hoppers and
one seaweed fly in her stomach when she died.

While some owl species are fairly restricted in their diet, most will take whatever is readily available. This sometimes results in seasonal or regional specialization in prey species that are abundant and easily obtainable. Prey choices can
vary even with time of day; burrowing owls, for example, catch small mammals
more often at night than during the day. Among the most opportunistic species
are the great horned owl and the eastern screech-owl.

Prey selection is influenced by the relative size of the owl and its quarry, but
large owls don't necessarily shun small prey. Some, like the great horned, will kill
almost any vertebrate or invertebrate they come across while hunting, up to the
maximum size that they can physically handle. Other large owls often kill small
prey by preference. Snowy owls, the heaviest owls in North America, eat almost
nothing but lemmings when these small mammals are plentiful, even though

Northern pygmy-owls will readily attack prey that outweigh them, but handling a hefty kill often proves challenging.

they are capable of killing animals as large as snowshoe hares and snow geese. North America's third-heaviest owl, the great gray, is a vole specialist in most parts of its range; it is equipped with relatively small feet and thin, needle-sharp claws that are well suited for catching these small rodents.

At the opposite end of the scale, pygmy-owls are renowned for their ability to kill birds and mammals that are bigger than themselves. These owls have proportionately long toes and talons to help them tackle large-bodied prey. The known prey of ferruginous pygmy-owls includes hispid cotton rats, which at 4 to 7 ounces (115–200 g) vastly outweigh even the heaviest of these 1.9- to 3-ounce (54–84 g) owls. And northern pygmy-owls have been seen attacking northern flickers, which are about triple their weight.

A northern pygmy-owl may not be able to overpower a flicker, but it can kill prey that are double its weight, as Thomas Balgooyen discovered when he caught a 1.8-ounce (52 g) northern pygmy-owl with a 4-ounce (119 g) immature California quail clutched in its talons. "When I picked them up," Balgooyen reported, "the owl seemed unconcerned and continued to grip and bite at the head of the quail during a 45-minute ride home"—a single-mindedness that is typical of this species when hunting. After weighing both birds, Balgooyen returned the still-alive quail to the owl and it finally completed its kill.

Pygmy-owls that catch oversized prey then face the challenge of transporting their prize. Sometimes they simply drag their meal to a sheltered spot where they can safely feed. One 2.7-ounce (76 g) northern pygmy-owl managed to get airborne while carrying a 1.9-ounce (55 g) vole, but it flew only 4 feet (1.2 m) off the ground.

Despite their propensity for killing large prey, captive northern pygmy-owls can happily survive on a single half-ounce (15 g) mouse per day. In the wild they probably need a bit more food, but not as much as would be provided by their largest kills. Rather than let the excess go to waste, they cache the uneaten remains and return to them later, a practice common to a number of owl species (see Prey Caching, below). Northern saw-whet owls can catch prey weighing up to about 1.4 ounce (40 g), but any that weigh more than 0.7 ounce (20 g) are usually consumed as two meals spaced at least four hours apart. In captivity these owls require about 0.6 ounce (18 g) of meat a day.

Moving up the size scale, great gray owls will eat up to seven 1.6-ounce (45 g) voles a day in winter. Snowy owls typically eat three to five adult lemmings or the equivalent every day, amounting to 5.3 to 12.3 ounces (150–350 g) of food daily throughout the year. A family of snowy owls consisting of two parents and nine offspring consumes approximately 1,900 to 2,600 lemmings during the course of one breeding season, but this hardly makes a dent in the total lemming population within their territory.

Hunting and Eating

Owls mostly use one of two basic strategies for hunting: aerial search or perch and pounce (also known as sit and wait). Aerial searching is best suited for wide open spaces, while the perch-and-pounce approach is most commonly used in treed habitats. In the absence of trees or shrubs, owls may perch on any somewhat elevated object or topographical feature.

Long-eared and short-eared owls are the only North American species that rely primarily on aerial-search hunting. Long-eared owls typically course back and forth about 1.5 to 6.5 feet (0.5–2 m) above the ground or sometimes cover the area in a pattern of overlapping circles. They fly with fast wingbeats interrupted by short glides, and often stall suddenly before dropping down to make a kill. Occasionally they hover. Short-eared owls make slow quartering flights, traveling about 1 to 10 feet (0.3–3 m) above the ground, or hover at heights of up to 100 feet (30 m). As they descend on their prey they often hover momentarily to correct their line of attack. Windy conditions sometimes prompt long-eared owls to switch to perch hunting, whereas short-eared owls are more likely to hover or hang in the wind.

Barn owls are predominantly aerial-search hunters, but often hunt from perches when they are less hungry or in habitats where this strategy is advantageous. When hunting on the wing they progress slowly at about 5 to 15 feet (1.5–4.5 m) above the ground. In their book *The Barn Owl,* British biologists D.S. Bunn, A.B. Warburton and R.D.S. Wilson evocatively describe this species' aerial hunting behavior: "Through binoculars one can see the turning of the head as the bird peers (*and listens*) intently into the grass, and sometimes one can even discern that the owl is following the progress of some small mammal.…It pauses, jinks sideways, hovers, retraces its 'steps', and thus responds to the telltale movements of its prey." When the owl finally strikes, its style is dictated by circumstances: "…it may dive forward, drop straight down with wings raised, throw itself sideways or appear to perform a somersault in the air to change its motion of flight by ninety degrees. Upon hitting the ground it lies where it falls, grasping at the prey with the feet and steadying itself by spreading its wings over the grass."

Snowy owls seem to be adept at both hunting strategies. When aerially hunting they fly back and forth methodically, frequently pausing to hover at heights of up to 50 feet (15 m). When the owl spots a lemming it either closes its wings and drops straight down on its prey or glides down gradually. In the treeless Arctic their perch hunting is done from rocks, hummocks or snowdrifts; on their southern sojourns they also use utility poles, fence posts and buildings.

Unlike other owls that live in open country, burrowing owls are not aerial searchers. Instead they hunt by pouncing from perches such as fence posts, by hovering and dropping (especially when targeting prey in tall vegetation), by flying up to catch prey in the air and by hopping, walking or running along the ground. A number of other owl species sometimes run on the ground in pursuit of terrestrial vertebrates or invertebrates, usually after having failed to grab hold of their prey on the initial strike from the air or from a perch.

Although a majority of owl species are primarily or exclusively perch-and-pounce hunters, they don't all go about things exactly the same way. Owls that rely heavily on hearing to determine prey location must use low perches to minimize the angle of error when pinpointing sounds in the vertical plane. Daytime hunters, on the other hand, often select high hunting perches that afford a wide view.

Northern hawk owls are highly diurnal and usually hunt from exposed, prominent perches, such as the tops of spruce trees overlooking meadows, marshes or old burns. When scanning for prey, a northern hawk owl leans forward on its perch, holding its body and tail almost horizontal, and sometimes pumps its

A sharp-eyed northern hawk owl watches for prey from a typical exposed perch.

tail. The moment it sights its prey, the owl's body becomes rigid. It then opens its wings and dives down into a fast, low glide that takes it straight to the target, sometimes with a few wing flaps to keep its momentum if the distance is great. One study conducted in Alaska's Denali National Park found that the striking distance for these owls ranged from 3 to 69 feet (1–21 m).

Boreal owls typically move from one low perch to another as they hunt, traveling through the forest in a zigzag pattern. They often watch for prey for less than five minutes at each perch, but if they detect a potential victim in an unassailable position, they may wait motionlessly for 10 minutes or longer, hoping it will make a wrong move. Distances between perches are short, generally less than 27 yards (25 m), and strikes are usually made within 11 yards (10 m) of a perch.

When northern pygmy-owls identify possible prey they often twitch their tail from side to side with a quick, jerky motion. Then, if they are not already in position to drop directly down or attack in a straight-line flight from their perch, they close in by moving from branch to branch until they are ready to strike.

Owls commonly use perches for hunting ground-level prey, but some species also go after prey in trees. Spotted owls in particular often pluck arboreal mammals, such as northern flying squirrels and red tree voles, from tree trunks or branches.

An elf owl delivers a cricket to its nest in a cactus. These mainly insectivorous owls use a wide range of hunting techniques.

Western screech-owls perch above creeks waiting for crayfish to move into the shallows, then swoop down and dip their feet into the water to seize one of these crustaceans. Eastern screech-owls catch fish, crayfish and tadpoles in the same manner, starting either from a low perch or from shore. Barred owls also fish from perches or wade in shallow water to catch aquatic prey.

Owls that prey on insects and other invertebrates on the ground, in the air or on plants may hunt from perches or use other specialized techniques. The elf owl's repertoire, which is characteristic of many such species, includes "hawking" (flying directly from a perch to catch insects on the wing), hovering before striking, running after prey on the ground, and flailing at vegetation to flush out hidden prey. Their diminutive stature also allows elf owls to probe for insects while walking on or hanging from flowers and foliage. Flammulated owls frequently hover and glean insects from tree trunks or conifer needles; they are so agile that they can invert themselves in flight, swinging their feet up to grab moths from below.

In contrast to the acrobatics of lightweight insectivorous owls, several species of large owls use their bulk for snow plunging—diving into deep snow to capture subsurface prey. Great gray owls, the masters of this technique, can punch through a snow crust solid enough to support a 175-pound (80 kg) person. In summer they sometimes plunge into the underground feeding tunnels of pocket gophers. Northern hawk owls, boreal owls and probably barred owls also snow plunge. The maximum recorded snow-plunging depth for a northern hawk owl is 12 inches (30 cm), while their greater body weight allows great gray owls to reach prey 18 inches (45 cm) below the surface, sometimes almost disappearing from sight in the process.

Excellent descriptions of snow plunging by great gray owls can be found in a classic book about this species by Robert W. Nero, one of North America's leading experts in great gray owls. "Typically, an owl would fly out from a perch, hover over the snow—apparently taking a final bearing on its prey—then partly fold its wings and plummet downward into the snow, sometimes vertically, at times at an angle," he writes. "The spectacle of these solemn-faced, relatively clumsy-looking owls sailing forth on their huge wings, then suddenly folding up and diving face downward into snowbanks astounded us."

By studying film footage and still photographs of great grays in action and examining plunge-holes left in the snow, Nero determined that plunging owls usually bring their feet forward under the chin just before hitting the snow, but sometimes they hold them well back beneath the tail and strike the snow face-first. The folded wings are held slightly out from the body, absorbing some of the shock of the impact and helping to brake the owl's momentum. The wings usually remain on or close to the surface, and the owls sometimes use their spread flight feathers to lift their bodies out of the snow. However, Nero notes, "at times plunging birds went so deep even the wrists went into the snow, so that only the tips of the primaries and tips of the tail feathers stuck up above the surface."

A plunging owl apparently does not have to connect with its target immediately. The force of its impact collapses the prey's snow tunnels and cuts off all escape routes, giving the hunter time to reach around until it can grasp the victim with its feet. The owl then lowers its head, seizes the mouse or vole in its bill and swallows it whole.

Owls have some hunting strategies that are rarely seen and may be practiced by only a few resourceful individuals. For example, some great horned and barn owls have been known to capture bats emerging from their cave in the evening by flying straight into the swarm with outthrust feet, presumably grabbing whichever animal their talons touch first.

In Ohio one winter, John James Audubon watched several snowy owls fishing in a curious manner that has never been reported by any other witness. In each instance the owl lay flat on a rock beside the water, lengthwise to the shore, and remained motionless until a fish rose to the surface nearby. The moment that happened, the owl "thrust out the foot next [to] the water, and, with the quickness of lightning, seized [the fish] and drew it out." It then moved some distance from the water to eat its catch before walking slowly back to the edge and resuming its supine hunting pose.

With talons spread wide, a great gray owl prepares to strike.

Another odd tactic was reported by A. Brazier Howell, who spent the winter of 1915–16 studying Arizona birdlife. According to the chicken rancher on whose property Howell camped, great horned owls were taking a heavy toll on his hapless birds. "He states that one will alight on a branch where a chicken is roosting," wrote Howell. "The latter will awaken and shriek, but is too scared to move. The owl then sidles along and grabs the fowl by the neck."

In most cases owls catch prey with their feet, using one foot or both, depending on the relative size of the hunter and its quarry. Elf owls and screech-owls are among the few North American species that sometimes use their bills, snapping up insects in flight.

When catching prey by the standard method, owls extend their legs with the feet close together and splay their toes to form a wide rectangle with their talons. Film footage of barn owls hunting shows that the owl "looks" straight at its prey as it approaches—even in full darkness, when it is actually positioning its head according to auditory cues. About 3 feet (1 m) from the target, the owl suddenly thrusts its feet forward until they almost touch its bill, then quickly pulls its head back and out of the way in time to strike with deadly precision.

Unlike hawks, eagles and falcons, whose toes are fixed in position with three pointing forward and one backward, owls have a reversible outer toe that can be pointed in either direction. They use this flexibility to maximize the catching area of their feet when making a strike, directing the middle two toes forward and the first and fourth toes backward. A great horned owl's fully spread talons cover an area of approximately 8 by 4 inches (20 by 10 cm). The barn owl's talon spread measures about 6 by 3 inches (15 by 7.5 cm), compared to just over 3 by 2 inches (8 by 6 cm) for long-eared owls, despite the fact that these two species are very similar in body size. The somewhat smaller burrowing owl has a 3-by-2-inch (7.5 by 5 cm) talon spread.

Owls, and other birds of prey, have powerful leg muscles that produce a strong grasp, as well as a specialized tendon-locking mechanism that allows them to sustain a forceful grip on their prey, which is useful for both killing and carrying. Once caught, vertebrates are killed either by the gripping action of the talons alone or by a skull-crushing or backbone-breaking bite with the bill, usually given while the animal is pinned to the ground or held firmly on a perch. In order to subdue snowshoe hares, snowy owls sometimes cling tightly and back-flap until the hare is exhausted, then deliver the *coup de grâce*. Invertebrates are dispatched in much the same way as vertebrates, or simply eaten alive.

Whether prey are eaten intact or piecemeal depends on their size and type and sometimes on the owl's appetite. When snowy owls are hungry they eat rodents whole; when less ravenous they bite or tear them into pieces. Great gray owls typically swallow small mammals whole and process larger prey by tearing off bite-sized chunks. Great horned owls devour small prey of all types in one bite and dismember larger prey, usually discarding the head and feet. Some small owls, such as eastern screech-owls and ferruginous pygmy-owls, partially pluck birds before consuming them in a piecemeal fashion, and may leave bird wings and tails uneaten.

Most owls habitually begin with the head when consuming vertebrate prey piece by piece, though some eviscerate the animal first. Similarly, when owls eat animals whole they nearly always ingest them headfirst. This inelegant procedure, as demonstrated by the barn owl, is well described by Bunn, Warburton and Wilson:

> the owl picks [the prey] up deliberately in the bill and manoeuvres it into a head-first position, often with the assistance of one foot. It then repeatedly opens the bill and jerks its own head forward to seize the animal farther along the body until it can be swallowed. If the prey is around maximum swallowing size the sequence is reversed once it has become lodged well inside the bill, the head jerking backwards and the bill and pharynx opening so that the momentum carries it into the oesophagus, leaving the owl looking rather uncomfortable for a while!

Screech-owls and elf owls engage in what one observer described as "finicky manipulations" when contending with large insects or invertebrates. They transfer the prey back and forth between bill and feet as they crush it and tear off

Regurgitated pellets, like these from a great horned owl, provide researchers with vital clues about owl feeding habits.

unappetizing or hazardous appendages, including wings, legs (which can choke the owl) and scorpion stingers. Soft-bodied large insects can be folded and swallowed whole, while other prey are held in a raised foot and eaten in bites. As with vertebrate prey, large invertebrates are usually eaten headfirst.

After finishing a meal, carnivorous owls often wipe their bill and face on a branch or other surface. Snowy owls often use snow for cleaning, and ferruginous pygmy-owls use tree bark.

Pellets

The disadvantage of eating prey whole or in large chunks is that each meal includes lots of indigestible body parts, such as bones, fur, feathers, claws, teeth, beaks, scales and exoskeletons, which have the potential to interfere with nutrient absorption and damage or block the digestive system. Owls deal with this problem by keeping these indigestible materials in the upper part of the digestive tract and compacting them into oblong pellets, which they subsequently regurgitate. Meanwhile they process and excrete other wastes in the usual manner.

Pellet production is not unique to owls. More than three hundred different bird species, including other birds of prey, gulls, herons, cormorants, kingfishers, crows and ravens, produce pellets. However, owl pellets contain the highest proportion of skeletal remains—approximately 10 times more than is found in pellets produced by hawks and other daytime raptors. This is because owls have a greater tendency to swallow intact prey, and the relatively low acidity of their gastric fluids limits their ability to digest bones and other hard materials.

When an owl consumes a meal, the food passes down its esophagus and ends up in the gizzard (also known as the muscular stomach or ventriculus). Here digestion proceeds as the walls of the gizzard repeatedly contract and relax, mixing the food with gastric juices, pumping liquefied nutrients into the small intestine and compacting the indigestible solids. X-rays of a great horned owl's gizzard, taken at hourly intervals after the owl polished off seven mice, showed that it took five hours for the forming pellet to attain a recognizable shape. By the 10th hour, compaction of the material was complete; the owl finally expelled the pellet 16 hours after eating.

The interval between eating and pellet-casting generally depends on the size of the meal. In another experiment, short-eared owls produced pellets within three hours of a 0.4-ounce (10 g) feeding, but not until eight hours after a 1.4-ounce (40 g) repast. Owls produce one pellet per feeding session, which may consist of a single animal swallowed whole, numerous small prey caught over a period of time or only a portion of a large kill. Once a pellet forms in the gizzard, the owl cannot eat additional food until it gets rid of the remains of its previous meal. Most owls produce one or two pellets every 24 hours.

According to Bunn, Warburton and Wilson, barn owls appear "decidedly miserable and listless" when they are about to rid themselves of a pellet. Other observers have noted that owls usually close their eyes and narrow their facial disk in a way that suggests significant discomfort. During regurgitation, which can take several minutes, an owl may lower its head or stretch its neck with its bill wide open. Often it will jerk or vigorously shake its head. Sometimes it repeatedly opens and closes its mouth. Muscular contractions of the upper digestive tract propel the pellet up the esophagus; when it reaches the mouth, the owl leans forward and drops it. Freshly ejected pellets are coated in mucus, which protects the owl's esophagus and mouth against the acidity of the gastric fluids in which the pellet has been immersed and also lubricates the passageway.

A knowledgeable person can often tell which owl species produced a particular pellet. Size is an obvious clue: the larger the owl, the larger its pellets. The pellets of great horned owls can be as much as 4 inches (10 cm) long, and those of great grays measure up to 3 inches (7.5 cm) in length, while pygmy-owl pellets are no more than 1 inch (2.5 cm) long. However, there are exceptions to the rule. The average length of a barn owl pellet is just under 2 inches (5 cm), but these owls occasionally produce pellets that are the size and shape of large marbles and consist of little but fur.

Owl researchers spend a lot of time looking for pellets and categorizing the various prey body parts they contain, because these reveal a wealth of information about feeding habits. Skulls and teeth are particularly useful for identifying mammalian prey species, while feet and feathers, as well as skulls, help distinguish bird species. Experts can classify invertebrate prey remains by examining jaws, legs or wing covers or even mere fragments of exoskeleton. However, soft-bodied invertebrates such as earthworms will be absent from the dietary record provided by a pellet. Pellet analysis can also underestimate the occurrence of some other types of prey. Pellets consisting solely of amphibian remains, for example, are rarely found because they disintegrate much faster than those containing fur or feathers.

In general, fresh pellets are darker, glossier and more solid than older ones, which tend to fade and fall apart as they age and are exposed to sun and rain or snow. However, the pellets of some species, such as the elf owl, have little integrity even when they are fresh. Large collections of pellets can accumulate where owls use habitual roosting sites or perches, while owls that roost randomly cast their pellets in widely scattered locations.

Prey Caching

During the breeding season, male owls of many species commonly kill and store prey in or near their nests. Prey caching during courtship may be a signal to a prospective mate that the male in question will be a good provider, or it may influence clutch size by indicating local prey abundance. During the nesting period, prey caching may help reinforce the pair bond, but it also clearly serves the practical purpose of ensuring there is sufficient food for the female during incubation and brooding and for the growing chicks.

Male snowy owls sometimes deposit uneaten lemmings by favorite perches at the beginning of the breeding season. Later, once the chicks have hatched, they cache extra prey both at their perches and at the nest, usually accumulating no more than five at a time.

Boreal owls start caching prey in the nest cavity one to two weeks before nesting begins and usually cease once the nestlings are about three weeks old. The peak caching period is from egg laying to hatching. Northern saw-whet owls also concentrate their prey storage efforts in this period, sometimes surrounding the incubating female with up to 24 offerings. With their larger nest cavities, barn owls can crowd in even more extra rations, often accumulating as many as 30 to 50 items. In what seems like true overkill, one barn owl nest in Michigan held 189 dead mice. Other owls are not quite as zealous in their provisioning, but researchers have found that food stored in elf owl and screech-owl nests sometimes ends up uneaten, perhaps because it exceeds the family's needs.

Prey caches in great horned owl nests during the early nestling period are distinguished by the size of the prey as well as quantity. One nest in the Yukon contained a dozen snowshoe hares, while another in Saskatchewan was stocked with 2 hares and 15 pocket gophers.

A snowshoe hare cached for future consumption (see the feet to the left side of owl) makes for a crowded nest, but guarantees that this female great horned owl and her young will not go hungry.

Short-eared owls continue their prey caching into the period when the young have left the nest but are still flightless and dependent on their parents. At this time both male and female short-eared owls stockpile prey, caching up to five items at once.

Burrowing owls store both vertebrate and invertebrate prey, caching them either in the nest burrow or in other burrows and tunnels throughout their hunting area, usually within 33 yards (30 m) of the nest burrow.

While prey caching is often associated with courtship and nesting, it also occurs under other circumstances. When an owl catches prey that is too big for one meal, it usually stores the leftovers and retrieves them later. And during inclement weather, especially in winter at northern latitudes, owls of certain species regularly kill and cache surplus prey as a safeguard against poor hunting conditions.

Pygmy-owls, with their proficiency in catching comparatively large prey, are among the species that frequently end up with uneaten portions of prey. Tree cavities are typical food storage places for these owls, but one northern pygmy-owl impaled a dead Bohemian waxwing on a blackberry thorn. In Texas ferruginous pygmy-owls sometimes store prey in patches of ball moss. Western and eastern screech-owls also commonly store small mammals and birds in cavities

In winter, boreal owls thaw frozen meals by warming the stored prey with their own body heat.

in winter, usually after beheading them. Eastern screech-owls cache prey in unconventional locations as well, such as the rafters of an old building.

Spotted owls are often wary when storing food, looking around frequently and sometimes moving the item several times before deciding on a final location. This might be on a moss-covered branch or tall stump or on the ground under a fallen log, in a crevice at the base of a tree or among moss-covered rocks. During hot weather they often select cache sites that are cooler than the surrounding forest. Once the owl is satisfied with its cache location, it usually sits upright and stares at the cached prey, then slowly backs away. Spotted owls also sometimes roost beside their caches.

In winter a northern hawk owl may capture and store up to 20 prey within a three-hour period. These owls cache food in crevices and in old woodpecker excavations in trees and stumps, on top of spruce branches and at the base of trees and fence posts. One individual used its bill to push snow over a kill that it cached below a tree, working until the prey was completely covered.

Another species that sometimes caches prey on the ground is the great horned owl, which has been known to hide dead snowshoe hares at the base of shrubs. Researchers in the Yukon Territory found that the owls rarely revisited these caches when hare population density was high, but usually returned to feed on the cached prey and sometimes even guarded it when hares were scarce.

Boreal owls and northern saw-whet owls typically store whole prey and uneaten portions of kills in tree-branch forks or on top of conifer branches. Boreal owls frequently roost close to their caches and will defend them against other birds, especially gray jays. When its cached prey is threatened, a boreal owl sits on the food, moves it to a new location or consumes it immediately.

Throughout much of Canada and the northern United States, cached prey quickly freezes in winter, becoming a solid mass that is too hard for an owl to dismember. Northern saw-whet owls and boreal and great horned owls, as well as perhaps others, solve this problem by thawing frozen meals with their own body heat. Biologist Søren Bondrup-Nielsen conducted a series of prey-thawing experiments with captive northern saw-whet owls. He learned that the owls were unable to consume mice that were colder than 30°F (–1°C) and generally preferred to warm the mice to 35°F (1.5°C) before starting to eat. To do this an owl grasped the mouse in both feet, holding it head-end forward, then sat down with its ruffled breast feathers over the frozen carcass, often completely concealing it. This posture, which was assumed by both sexes during prey thawing, resembles the position females use when incubating eggs.

During the thawing process Bondrup-Nielsen's owls periodically stood up, still holding their prey, and bent over to bite the mouse's head and neck, apparently testing to see whether it was "done." If it was still too solid they sat back down and continued warming. It took the owls 20 to 25 minutes to sufficiently thaw mice that were offered to them while frozen to 3.2°F (–16°C) and from 2 to 10 minutes for mice whose initial temperature was just below 30°F (–1°C). The owls often stopped the thawing process as soon as the mouse's head and upper body were edible. After eating those parts they either resumed thawing the mouse or, if they were satiated, allowed it to refreeze, then rewarmed and ate the rest later.

Prey thawing probably draws substantially on an owl's energy reserves, but it is an effective means of avoiding starvation in winter. At least some species of owls are also able to endure periodic food shortages despite severe weight loss. A whiskered screech-owl and an elf owl that were experimentally deprived of food for three to four days each lost about a quarter of their body weight but suffered no other apparent ill effects. There is also some evidence that burrowing owls can survive an enforced fast when trapped underground for several days by heavy snow.

Drinking

Little is known about how much water owls drink. Some species, such as the great horned owl, obtain all the water they need from their food. Others require supplementary water. Burrowing owls, screech-owls and pygmy-owls are among the species that have been seen drinking in the wild, from sources such as natural springs and shallow creeks. Ferruginous pygmy-owls will drink from backyard bird baths, sometimes on a daily basis. Northern hawk owls sometimes consume snow after eating prey.

CHAPTER FOUR

Communication

Having learned as children that owls hoot, many people grow up unaware that this is not true of all owls or that their repertoire may also include whistles, screams, moans, shrieks, barks and more. As Henry David Thoreau wrote after listening to an eastern screech-owl,

It is no honest and blunt *tu-whit, tu-who* of the poets, but, without jesting, a most solemn, graveyard ditty, the mutual consolations of suicide lovers remembering the pangs and the delights of supernal love in the infernal groves....*Oh-o-o-o-o that I had never been born-r-r-r-r-n* sighs one on this side of the pond....Then—*that I never had been bor-r-r-r-n* echoes another on the further side with tremulous sincerity, and *bor-r-r-r-n* comes faintly from far in Lincoln woods.

Clearly this is not the clichéd hoot-owl of childhood picture books!

A nighttime existence favors acoustical communication over visual signals, but a number of owls have developed visual displays that complement or even take precedence over vocalizations. Visual displays are most often performed by species that are active during daylight or twilight hours. Together or separately, vocalizations, mechanical sounds (wing clapping and bill snapping) and displays impart a wide range of information to family members, competitors, intruders and enemies.

A three-month-old eastern screech-owl tries out its voice, while its sibling listens.

Songs, Calls and Other Vocalizations

Owls, especially those that are mainly nocturnal, rely heavily on their voices for communicating. Like other birds, they use songs and calls for a wide variety of functions, many of them associated with reproduction. Each owl species has a unique vocalization, often referred to as the primary song, which is used for territorial proclamation, nest-site advertising and mate attraction. A few also have a secondary song that serves some of these purposes. Although these vocalizations are often not particularly melodious or elaborate, they are songs in the ornithological sense of the term because they are associated with reproduction and consist of a set configuration of notes and phrases repeated in a formulaic manner.

In many owl species the primary song or territorial advertising call is given exclusively or predominantly by males. When females share this part of the vocal repertoire, they typically give these songs or calls less frequently than their mates and in somewhat different contexts. The repertoire of most species also includes female-only calls that are associated with courtship and nesting. Because males typically establish and reinforce pair bonds by presenting prey to the female, females commonly utter food-begging calls when interacting with their mates.

Each species also has an assortment of other vocalizations that may or may not be gender-specific, such as contact calls and alarm calls. Generally calls are acoustically simpler than songs and have no standardized structure. Since owls mostly vocalize at night, researchers are often uncertain about the meanings of their calls and may not even be able to determine the sex of the caller. Variability in how, when and where certain calls are delivered and gradations between some types of calls also complicate matters.

Within the Strigidae family the female's syrinx (voice box) is almost always smaller than the male's. As a result, female vocalizations are higher-pitched than those of males. In the barn owl the female's syrinx is larger than the male's. This anatomical difference accounts for the fact that the female's screech has a huskier tone than her male counterpart's.

Depending on the species, sexual differences in pitch may or may not be clearly audible. Paired great horned owls often synchronize delivery of their territorial hooting, providing listeners with an excellent opportunity to compare male and female voices. During these double acts the female begins by delivering a three-second series of six or seven hoots and the male responds with a five-hoot song, which also lasts about three seconds. The male's song overlaps his mate's or starts within seconds after hers ends. Each pair of songs is followed by a 15- to 20-second pause. A duet session usually lasts at least 10 minutes and can go on for up to an hour.

Several other North American species also sing pair duets or engage in call-and-response exchanges, alternating between the male's primary song and the female's courtship or mate-contact call. These include the mottled and long-eared owls and all of the screech-owls. Differences in voice pitch can help human listeners distinguish pair duets from duels between neighboring males. With eastern screech-owls the pattern of the singing provides another clue. Paired males and females of this species sing in turns, performing their duets both day and night throughout the incubation and brooding periods. Rival males, on the other hand, sing synchronously and only at night.

Sexual differences aside, the size of an owl's syrinx—and therefore the pitch of its calls—generally corresponds to body size: the larger the owl species, the lower-pitched its voice. The two exceptions in North America are the flammulated and mottled owls. Both have voice boxes that are unusually large for their body size and more specialized than the voice boxes of other owls. Consequently their hoots are strikingly low in pitch compared to other owls of a similar size.

Young owls begin vocalizing shortly before hatching or soon afterward. Their initial calls are generally simple peeps, squeaks or twitters, which become progressively louder with age and are gradually replaced by species-specific nestling vocalizations. These nestling vocalizations, as well as the fledgling vocalizations, are often recognizable precursors of adult calls. While the vocal development of each species is different, the following examples illustrate the general pattern.

During their first month of life, barn owls chitter frequently when experiencing discomfort or seeking parental attention, as well as when quarreling with their siblings or exploring the nest cavity. As they get older they chitter less, and by the time they fledge the nestling chitter has been replaced by adult-style chirrups, twitters and squeaks. Another characteristic call of barn owl nestlings is a widely variable (raspy, wheezy or whistled) hiss, commonly referred to as the snore call. The snores of newly hatched barn owls are weak and infrequent, but from the time they are two weeks old until they fledge, nestlings snore very loudly and persistently, especially when hungry. Adult females snore often during the breeding season, when soliciting food and interacting with their mates. Adult males also snore, but only occasionally.

Besides peeping and twittering, young elf owl nestlings utter rasping notes at a rate of up to 48 times a minute, increasing the volume in relation to their level of hunger. In the open desert, rasps coming from an elf owl nest cavity can be heard from 330 feet (100 m) away. After being fed, the owlets give high-pitched trills that closely resemble the adult female's food-solicitation call. Older nestlings and fledglings give poorly developed versions of the puppy-like yipping notes that make up the adult male's primary song.

Young burrowing owls give two types of rasping calls. The milder version is a hunger call, given before and during meals. The other, a more intense and prolonged vocalization, sounds remarkably like the rattling buzz of an agitated western rattlesnake—a resemblance that is particularly startling when the call is given from inside a nest burrow. Acoustic analysis of the owl's rattlesnake rasp call and of actual snake rattling has confirmed this similarity, revealing that the two sounds are structurally similar: both are wide-frequency noises with their energy concentrated in much the same range of frequencies. This vocal mimicry may well have evolved to deter predators such as badgers and weasels from entering nest burrows. It has definitely proven effective on occasion in discouraging ornithologists from reaching blindly into burrows.

Vocal development in owls is innate, rather than learned, meaning that even an owl raised in isolation will produce the same calls and songs as the rest of its kind. The rate of vocal development is consistent for all members of a species, but it varies between species. From the time they are about three weeks old, eastern screech-owls can produce all of the vocalizations in this species' adult repertoire, except for the two songs. They gain these after they fledge, but sing infrequently prior to their first breeding season.

In comparison, juvenile male great horned owls make preliminary attempts at adult hooting during their first winter, but their initial hoots come out mostly as gasps, gurgles and squawks. They rarely master complete sequences of mature deep-toned hoots until springtime. Juvenile females appear to skip the trial phase; instead they suddenly start producing fully developed female-type territorial hooting during their first spring. Within great horned owl populations the frequency and length of the syllables in male hooting sequences vary from owl to owl, but for any given male these elements are consistent over time. As a result, each male's distinctive hooting song identifies him individually, as well as asserting his territorial ownership and inviting female company.

Individualization of primary songs has also been detected in some other owl species. Northern saw-whet owls individualize their primary song with subtle variation in frequency. In one study the average frequency of five different males' advertising calls, recorded over four consecutive nights, ranged from 1007 to 1210 hertz. Presumably owls can perceive distinctive song characteristics of this type by ear, but humans generally have to rely on analysis of sonograms (graphic representations of sound recordings).

Spotted owls can apparently imitate the nuances of their neighbors' calls, a talent that is thought to be rare among owls. Although spotted owls are highly territorial, disputes between adjacent territory-holders are rare, perhaps because of this species' well-developed song-recognition abilities.

During the breeding season, snowy owls are highly vocal around the clock.

The best time of year to hear owls sing their primary song depends on the species. Migratory or nomadic owls do most of their singing in late winter and early spring, while establishing territories and forming pair bonds. This tendency is well illustrated by the snowy owl, a species that is largely silent outside of the breeding season. On the arctic nesting grounds, however, males fill the air with their booming hoots from April to August, especially during the earlier months.

In contrast, there are often two peak singing periods in species that remain on their territories year-round—one at the beginning of the breeding season and one in late summer or fall. The latter is probably tied to juvenile dispersal and the need for adult territory-holders to defend their tenure.

Although singing generally slows down once egg laying begins, it doesn't cease entirely, particularly when unmated males continue to attempt to attract females. For example, male flammulated owls sing for hours on end in the early part of the breeding season, but only sporadically and late at night once they have hungry youngsters to feed. Meanwhile, their bachelor neighbors sing vigorously every night throughout the summer.

Flammulated owls are not the only ones that engage in prolonged singing sessions. Among the more energetic singers are long-eared owls, which sometimes sing for up to five hours at a stretch. One western screech-owl was monitored as it sang continuously for two hours, repeating his bouncing-ball song eight times a minute, for a total of 960 notes. Elf owls often sing almost without interruption

for an hour or more, and, like many owls, they increase their rate of singing on clear nights under a full to half moon. A session of advertising calling by a male northern hawk owl, on the other hand, may last only a quarter of an hour, with intervals of two to five minutes between each call.

The time of night or day when owls are most vocal generally reflects species activity patterns. Highly nocturnal owls are seldom heard when it is not dark. Others may sing or call at any time of night or, occasionally, during the day. The territorial singing of species that are most active in the twilight hours typically begins around dusk, then tapers off late at night, with a second peak of singing in the pre-dawn hours. Male northern hawk owls often give their advertising calls in early morning and evening, despite being active throughout the day.

As with primary songs, the incidence of other vocalizations varies seasonally and is tied to daily activity patterns. Alarm calling is relatively infrequent during incubation and reaches its height when parents are caring for vulnerable nestlings and fledglings. Calls associated with prey deliveries increase along with the feeding demands of growing chicks. These include calls by food-bearing males announcing their arrival and food-begging calls given by brooding females and nestlings.

Once the young leave the nest, their vociferous petitions for food may continue for weeks or months, depending on the species, and parents and offspring may also exchange frequent contact calls. Fledgling great gray and great horned owls give very similar strident shrieks when hungry, while the food-begging calls of juvenile long-eared owls sound like a squeaky gate or trees creaking in the wind. Biologist David Parmalee, who has spent years studying snowy owls on their breeding grounds, describes the call made by young snowy owls that have dispersed from the nest as "an ear-splitting squeal." It is unnerving, he says, "to hear these rending screams issuing from a multitude of hiding places, and not to know their source."

Primary songs and calls are generally given from characteristic locations, frequently at or close to the nest. Male boreal owls, for example, sing from tree perches that are usually within 330 feet (100 m) of the nest, and often much closer. Northern hawk owls give their advertising calls from prominent perches or during display flights.

Male northern saw-whet owls mostly broadcast their advertising song from a high but hidden perch such as a branch within the crown of a tree, but occasionally from a potential nest cavity. Male flammulated owls also select concealed singing perches, typically next to a tree trunk or within a dense clump of foliage. Some other cavity-nesters, such as the elf owl and screech-owls, sing more frequently from inside prospective nest holes or while perching in the cavity entrance. When singing their primary song outside the nest cavity, whiskered

Northern saw-whet owls usually sing from concealed perches, high in the tree canopy.

screech-owls often use an exposed perch such as a bare branch, and habitually return to the same site.

Owls that breed in open country ordinarily vocalize while on the ground or in flight, since perches are few and far between. Male burrowing owls often sing at the mouth of their chosen nest burrow. Male snowy owls perform their territorial hooting from favorite prominences or while flying. Male short-eared owls usually sing their primary song during courtship flights, but may also sing from the ground or an elevated perch.

Other vocalizations may also be delivered from locations that are typical of the species, and sometimes specific to an individual owl. Female ferruginous pygmy-owls, for example, adopt a habitual perch near their nest from which they give both food-begging and alarm calls. Northern hawk owls are among those that typically give alarm calls on the wing.

Larger owl species can generally project their calls farther than smaller ones, though there is not always a direct relationship between an owl's size and the number of decibels it puts out. Hooting snowy owls have been heard by humans at distances of up to 7 miles (11 km), but even under ideal conditions the great gray owl's primary song is not audible beyond about half a mile (800 m). Boreal owls have powerful voices for their size. Their primary song is often perceptible a mile (1.6 km) away and can sometimes be heard at a distance of more than 2 miles (3.2 km).

Because of their intended purpose, territorial and advertising songs or calls are usually louder than most other vocalizations, with the exception of alarm calls. For example, the male long-eared owl's primary song can carry more than half a mile (800 m), whereas the female's nest call—a courtship and contact call that has been likened to the sound made by blowing through a paper-covered comb—is barely audible 200 feet (61 m) from the nest.

Habitat and weather also influence audibility. One researcher found that he had to be within 330 yards (300 m) of a northern saw-whet owl to hear it singing in forested habitat, whereas he could hear the same song at triple the distance when it was traveling across water. In the forest on a calm night, a flammulated owl's hoots are perceptible from 550 yards (500 m) away, but they may travel twice as far when carried on a breeze.

Bill Snaps and Wing Claps

Two of the sounds that owls use to communicate are produced mechanically rather than vocally. The most common of these is bill snapping, a behavior apparently exhibited by all species of owls. Both nestlings and adults bill snap, typically beginning within a few weeks of hatching. In some species bill snapping seems to be mainly an expression of stress or discontent in threatening situations, while in others it appears to also serve an aggressive function. Burrowing owls, for example, often bill snap in conjunction with defensive vocalizations such as rattlesnake rasps and screams. Other species, including the boreal owl, often hiss in association with bill snapping. Barn owls are one of the few species that sometimes bill snap during courtship displays as well as when threatened, but members of this species do not bill snap as loudly or as frequently as many other owls.

Owls regularly bill snap when being handled by humans, so biologists have had plenty of opportunity to watch them in action, but there is still some uncertainty as to how they make this sharp clacking or popping sound. Although most observers believe it is produced by the owl clapping its mandibles (the upper and lower halves of the bill), some suggest it is made by clicking the tongue, or that both actions are involved.

Wing clapping is restricted to a minority of species, most of them in the genus *Asio*. This sound is made when a flying owl brings its extended wings together abruptly, usually below the body. There is one published report of a barn owl clapping its wings above its body, but this is the only record of upward wing clapping in this or any other owl species.

The wing clapping of short-eared owls sounds like a person slapping hands against thighs in an alternating rhythm and as rapidly as possible. Both sexes use defensive wing clapping to drive potential enemies away from the nest, and males wing clap during their elaborate courtship flights, giving a burst of multiple claps at two different points in the performance. Male long-eared owls typically wing clap from six to twenty times during their courtship flight but, unlike the short-eared owl, they give the claps singly and at irregular intervals. Female long-eared owls wing clap occasionally.

Wing clapping is also characteristic of the closely related stygian owl, but there are few published reports of this behavior. John Arvin, an authority on Mexican birds, compares this species' wing claps to the sound of a .22 rifle being fired overhead. For several years he tried calling in male stygian owls by playing a voice tape, and invariably he heard the same loud, sharp cracks in response.

Male northern hawk owls and barn owls also wing clap. Like *Asio* owls, they make these sounds as part of their aerial courtship displays. Barn owls wing clap infrequently and fairly quietly, executing either single claps or one loud clap followed by a softer one.

In a typical display of ritualized feeding, a male great gray owl in Idaho presents a pocket gopher to his mate.

Visual Displays

Owl displays generally fall into two categories: those directed at confirmed or prospective mates, and those directed at potential predators and other threatening beings, including humans.

Ritualized feeding, in which the male presents prey offerings to the female, seems to be part of the lead-up to mating for all owls. Mutual preening and nest showing are also common aspects of pair bonding and preludes to copulation. The latter behavior typically involves the male flying back and forth to a potential nest site—and, if it's a cavity, popping in and out of the nest hole—as he invites the female to inspect the site and give it her seal of approval. Beyond these rudimentary displays, owls of most species rely primarily on their voice to make their intentions known to prospective mates. In North America, owls that do little or no displaying include the spotted, barred, boreal, northern saw-whet, flammulated and elf owls and the pygmy-owls and screech-owls.

Several species, however, have showy courtship and pre-copulation displays. One of the most impressive is the short-eared owl's sky dance, which males perform night and day, starting in late winter. The male begins his aerial acrobatics by flying in tight circles as he ascends to a height of 200 to 300 feet (61–91 m), then drops a short distance while wing clapping eight to twelve times in quick succession. Again he circles and climbs, this time to a height of 100 to 500 feet

(30–152 m), where he pauses and sings his courtship song—a rapid series of 16 to 18 tooting notes—as he hangs on the wind or flies forward with wings and tail fanned. He then drops into a shallow stoop, clapping his wings five to ten times as he descends. Throughout this performance the female usually watches from a ground-level perch, giving intermittent *keeeyup* calls. Occasionally she joins the male as he sky dances.

Finally, after repeating the entire sequence several times, the male spreads his wings wide at a slightly upward angle and glides down toward the female, rocking from side to side in the air. As he swoops past the female she sometimes launches into a chase, during which the two birds may use their talons to grapple with one another. Otherwise she follows him into the grass, where biologists assume that the pair copulates, though few people have witnessed this finale.

The male long-eared owl's courtship display, which is performed between dusk and dawn, is a zigzagging flight that takes him around and through the nest grove. It can also involve circling above the nest. Displaying owls fly with deep wingbeats interrupted by long glides and single wing claps. In the male's pre-copulatory display he perches near the female or on the nest and begins swaying while raising and lowering his wings. The female responds by crouching with drooped wings, usually on a branch but sometimes on the ground. One of the two birds then flies to the other and the male briefly mounts the female as they both bend forward with their wings extended. Sometimes they engage in mutual preening immediately before or after copulating.

Much of the northern hawk owl's courtship behavior is based on vocalizations, even though this species is active mainly in the daytime. The most conspicuous visual component is the male's circling display flight. Such flights are characterized by frequent gliding above the treetops and intermittent wing clapping while flying among the trees. During the glides the owl's extended wings are held stiffly and his head is elevated. Before copulating, paired northern hawk owls usually engage in loud trilling duets, sometimes bowing and touching their foreheads together as they exchange calls.

Barn owl courtship features several types of aerial display. During sexual chases the male persistently follows the female as she twists and weaves, often traveling at high speeds. Although chases are sometimes silent, these flights are usually accompanied by frequent screeches from both birds. In the "moth flight" the male hovers for several seconds in front of the perched female with his legs dangling. Wing claps are occasionally heard during this display. The in-and-out flight is a fast-paced display in which the male swoops into a proposed nest hole and calls from inside, then flies back out and screeches as he circles around, repeating this sequence until the female is enticed to join him inside the nest cavity.

The male snowy owl's courtship display has both an aerial and a ground component. The display begins with an undulating flight past the female, who is usually perched but sometimes airborne. The male flies in an exaggerated manner, raising his wings high above his back and bringing them close together under his body. Each pronounced upstroke is followed by a brief pause and a slight loss of altitude, which is quickly regained on the powerful downstroke. At the end of the flight the male ascends gradually, then makes a gentle vertical descent with his wings either flapping or held in a V above his back, and lands on an exposed patch of ground, frequently an old nest site.

A courting male snowy owl often carries a dead lemming during his display flight, usually in his bill or sometimes in his feet. On rare occasions the male transfers the prey to his prospective mate in midair, the two birds swinging gracefully upward so that their feet come together and the female can grasp the lemming with her talons as the male lets go. Typically, however, the male drops his prey at some point during the ground display, which commences upon landing.

The ground component of the display usually lasts one to two minutes, less commonly extending for up to five minutes. It starts with the male in a fairly erect posture with his wings partly spread and wrists raised high. Turning in place or walking about, he slowly orients himself to face away from the female so that she is either behind him or to one side. As the display continues the male leans farther and farther forward with his head lowered and his tail partly fanned, until he is more or less prostrate. Meanwhile the female usually flies or hops closer.

The male snowy owl also strikes an exaggerated pose when engaged in bouts of territorial hooting. He raises his head, swells out his throat enormously, lifts his tail until it is almost vertical and emits several long, low, sonorous hoots, bowing forcefully with every note. Mutual hooting sessions can involve as many as a dozen owls on neighboring territories, their calls reverberating across the tundra.

The great horned owl's courtship display has a number of elements that are similar to the snowy owl's—not surprisingly, since the two are in the same genus—but the former does all its courting while perched in trees. The male great horned owl initiates the performance by cocking his tail, puffing out his bib of white neck feathers and delivering a series of hoots, along with much bowing, tail bobbing and jerking. Emphasizing the white bib may enhance the display's visual impact in the low-light conditions under which the display is performed.

The female's answering hoots encourage the male to move closer, and the pair then carries on their duet, bowing with wings drooped or held out at an angle. Bill snapping, screams and high-pitched calls are also part of the mix. Both birds expand their bibs and bob their tails in unison as they hoot, but the male's actions are more pronounced. Courtship duets usually lead to a sexual encounter, which occurs after the female flies to the pair's copulation perch and is followed by the male. One pair, observed copulating on 12 different occasions, hooted from 13 to 196 times during their preludes, calling at a rate of two to three times a minute. During their brief copulation they both hooted four to five times a second.

The courtship and pre-copulatory behaviors of burrowing owls feature an assortment of displays, vocalizations and postures by both sexes, including leg and wing stretches, preening and prey presentations. While most of this activity occurs on the ground within 16 yards (15 m) of the nest-burrow entrance, some of it is aerial. In one type of display, performed mainly by males, the owl flies in wide circles. In another the displaying owl repeatedly makes a rapid ascent to approximately 100 feet (30 m) above the ground, hovers for five to ten seconds, then descends swiftly about halfway down before starting to climb again.

Male burrowing owls perform a characteristic bowing display when singing their primary song. From a standing position the owl bends forward, bringing his body nearly parallel to the ground and holding his wings together over his back, with the undersides of the flight feathers touching. While he is bowing, the white patches on the throat and brow are shown to their fullest extent. Since burrowing owls begin singing at sunset and continue through the night, these white facial markings likely increase the visibility of the display.

The white throat and brow patches are also highlighted in the burrowing owl's "white-and-tall" stance, which is part of the pre- and post-copulatory behavior of both sexes. Immediately before copulation the male stands upright, with his white facial markings exposed and his body feathers raised, and looks at the female from a short distance away. She assumes the same posture and shows her white facial patches, but doesn't stand as tall or raise her body feathers. He then flies to the female and they mate while giving characteristic copulation calls. After their sexual encounter is over, the male usually returns momentarily

The spread-wing posture demonstrated by this juvenile great horned owl is a common threat response.

to the white-and-tall stance, but without his body feathers raised, then flies back to the spot where he was previously singing. The female may or may not repeat the stance.

Great gray owls may perform one of the more unusual owl courtship displays. According to biologist Robert Nero, these owls sometimes execute late-winter snow plunges that seem to be more demonstrative than functional, at least in terms of capturing rodents. On several occasions Nero witnessed "a frenzy of plunging activity" and thought that the "overly desperate plunges suggested something more than hunger as motivation." This pseudo-hunting was distinguished from true hunting by the "casual, random fashion" in which the owls repeatedly dove into the snow, with hardly enough time to have detected any hidden prey and with little apparent effort to actually catch anything.

Owls respond to threats to themselves and their nests, eggs or young with a range of behaviors, all aimed at intimidating and repelling the foe in situations where fleeing is not an option. If initial communications fail to get the message across, they intensify their efforts and finally attack. Threat responses invariably include alarm calls, other defensive vocalizations and bill snapping, and many species augment their calls and bill snaps with visual displays. Some owls, such as the flammulated, have no known visual threat display.

Owl threat displays typically feature actions such as ruffling the body-contour feathers to create an illusion of greater size, lowering the head, drooping the wings and rotating them forward, crouching, and swaying from side to side. An additional characteristic of the barn owl's display is bouts of rapid head shaking, which punctuate their regular swaying and head weaving at intervals of about 10 seconds.

The adults and young of a number of species—including great horned, short-eared, long-eared and barn owls—further increase their apparent size when displaying by adopting a spread-wing posture. In this posture the owl raises its wings and fans them out above and to the sides of its head. By turning the leading edges of the wings downward and angling the trailing edges upward, the owl brings the upper surfaces of the wings into a forward-facing position.

The escalating threat displays of the great horned owl typify the behavioral progression shown by many owl species. Both adult and young great horned owls perform these displays, which grade into one another. Mild threats elicit bill snapping, hissing and occasional low, drawn-out screams and guttural calls. Second-level displays include the spread-wing posture as well as more agitated bill snapping, screams and other calls. In the most extreme form of threat display the owl hisses and utters higher-pitched, more prolonged screams while positioning itself so it is poised to strike with its feet. Beyond this point, attack replaces display as the owl hops forward and uses its talons to grasp at and rake its adversary.

Sometimes owls use a distraction display to try to draw potential predators away from their nest, especially after hatching, or away from young that have left the nest but are not yet capable fliers. Among the species that perform distraction displays are the barred, great gray, great horned, snowy, long-eared and short-eared owls. The usual form of display is a broken-wing or wounded-bird act.

The short-eared owl's distraction display includes precipitous drops or tumbles through the air, bringing the displaying bird to the ground some distance from the nest. There the act continues, with much flapping of outstretched wings and squealing cries. If approached, the owl slowly folds its wings and takes flight, sailing a little farther away and repeating the routine.

Snowy owls stage even more compelling performances, as ably described by David Parmalee. From a classic threat-display posture the owl "suddenly goes limp" and changes tactics. "Like some miserable, tortured clown, it scuds along, flailing turf and snow with great wings, stopping momentarily to rock and wallow, all the while uttering high, thin squeals interspersed with weird squawks." Every now and then the owl abruptly stands up and looks around or flies a short distance ahead of the intruder before resuming its "wing-flopping and wallowing." One determined female that Parmalee followed continued her act across 2 miles (3.2 km) of open ground to lead him away from the spot where she had left her offspring.

CHAPTER FIVE

The Mating Game

Owls have long breeding seasons compared to the majority of familiar daytime birds. Some North American owls feel their first hormonal stirrings as early as midwinter, and by the time spring loosens winter's grip the mating game is well underway for most species. The parental duties that follow successful courtship and breeding are generally not completed until late summer or fall.

Their singing and calling make owls conspicuous during the breeding season, but the average person seldom gets a chance to directly observe courtship, pair bonding or sexual activities. Sometimes, however, all it takes is patience and a good flashlight to gain a glimpse into this fascinating part of owl life.

Mating Strategies

Because of the extensive parental involvement of both sexes, monogamy—either seasonal or long-term—is the standard mating strategy for owls. But in the bird world monogamy isn't always what it seems. Among songbirds, genuine monogamy is the exception rather than the rule. In many such species females regularly have some or all of their eggs fertilized by males who are not their apparent mates. Among owls, on the other hand, females almost never copulate with males other than their mate. DNA fingerprinting of nestling owls shows that they are nearly always the genetic offspring of their putative father. This is true even of species that nest semi-colonially and presumably have more opportunities for philandering, such as flammulated, long-eared and burrowing owls.

With a prey offering in its bill, a male great gray owl courts a female at dusk.

Owls that hold year-round territories tend to mate for life, or at least for several years in a row. Included in this category are great horned, barred, spotted and barn owls, western screech-owls and ferruginous pygmy-owls. Seasonal monogamy is common among owls that are migratory or that wander in winter, since mates are not bound by territorial attachments through the non-breeding season. However, flammulated owls mate for life, despite their migratory lifestyle. Each spring, if both birds have survived the winter, they return to their previous year's territory, where they reunite and renew their pair bond. If the female fails to show up, the male attracts a new mate to share his old territory. If the female is the sole survivor, she finds a vacancy with a lone male on an adjacent territory.

In some species the duration of the pair bond depends on where the birds live. Migratory burrowing owls generally change partners from year to year, whereas those that belong to resident populations mostly mate for life. In Oregon, Idaho and California, the pair bonds of great gray owls appear to endure until one of the pair members dies. However, great gray owls that inhabit the boreal forest regions of Canada and Alaska do not maintain pair bonds through the winter, and they only remate with the same partner in successive years if prey populations remain high enough to keep both birds in the area.

Polygamy is rare among owls. When it does occur it usually takes the form of multiple-nest polygyny: one male mates concurrently with two or more females. Polygynous mating has been confirmed in 10 species of owls worldwide and is suspected in a couple of others. North American species for which there are documented cases of polygyny are the eastern screech-owl and the barn, boreal, northern saw-whet, long-eared, burrowing, snowy and northern hawk owls. Possible instances have been reported for great gray and elf owls.

Male owls probably provide more parental care in terms of providing food for their mates and offspring than any other birds except hornbills. This responsibility means they can try to support a second family only when food supplies are particularly abundant. Periodic population explosions of prey such as voles, lemmings or snowshoe hares often set the stage for polygyny, but even under these conditions only a small percentage of the males in any population will attempt this mating strategy.

For cavity-nesters polygyny also depends on an abundance of nest sites. Most, if not all, polygynous matings by northern saw-whet and boreal owls depend on nest boxes to augment the availability of natural nest cavities. On two occasions males of these species (a boreal owl in Sweden and a northern saw-whet owl in Idaho) have even mated with three females at once, all of which used nest boxes. Under natural conditions, triple polygyny by owls is completely unknown.

A polygynous male owl typically begins the breeding season with only one mate, which he feeds generously before and during egg-laying. Once her clutch is complete he starts singing near a second nest site. If he succeeds in attracting another mate, the male feeds her consistently throughout the egg-laying period while continuing to provide for his primary mate. Once the second clutch has been laid, the male focuses on feeding the first female and her brood, attending to the second family only when he has met the first family's needs. Consequently, nestling mortality in the second nest is usually higher than normal, and many secondary females end up raising few or no offspring to fledgling age.

Because polygynous males' nests are usually well separated from each other, most secondary females may be unaware of their mate's divided loyalties. However, female deception is not necessarily requisite for polygyny. When nest sites or bachelor males are in short supply, females—rather than not breeding at all that year—may knowingly take a chance on a male that is already mated.

When they share a nest site as well as a mate, there's no doubt that females are actively choosing polygyny. Only two instances of successful same-nest polygyny have ever been documented among owls; this exceedingly rare situation may be restricted to the barn owl, a species known for its highly flexible reproductive behavior. Surprisingly, the females in both cases showed no aggression toward their co-nesters. Their nest boxes were roomy enough that they could have put some distance between themselves, but they chose to incubate side by side, with only an inch or so between the two clutches, and they continued to accept each other's presence throughout the nestling period. In the same study area in Utah where these females shared nests, two other attempts at same-nest polygyny failed. In one, all three members of the trio disappeared after two eggs had been laid. In the other, one of the females (who may or may not have laid any eggs) departed during egg-laying or incubation, leaving the other female and the male in a monogamous relationship.

In the case of the three-timing male northern saw-whet owl in Idaho, at least two of the females must have known the score because of the close proximity of their nest boxes: two were only 16 yards (15 m) apart and the third was 143 yards (130 m) away. One of the females failed to hatch any of her eggs, but the other two each produced five fledglings—as many as or more than any of their monogamous neighbors. For the male it was a particularly successful breeding season: he fathered ten fledglings, while monogamous males who nested nearby fathered from two to five.

A study of snowy owls on Baffin Island during a summer when lemmings were at the peak of their population cycle also showed how profitable polygyny can be for males. By breeding with two females, one male fathered fifteen young, all of which fledged. Meanwhile, none of the monogamous pairs on adjacent territories raised more than nine fledglings, and some raised as few as seven.

Female owls, like this snowy owl, have full responsibility for incubation and brooding. Males are the principal providers of food.

Among owls, incubation and brooding are entirely female responsibilities, so it is impossible for a female to divide her efforts between two males and lay eggs in each of their nests at the same time, a strategy used by some other birds. But when prey populations are high, female owls occasionally abandon their offspring late in the nestling period and start a second family with another male. This breeding strategy is known as serial polyandry. Boreal, long-eared and barn owls are the only owls for which serial polyandry has been confirmed, but biologists suspect that northern saw-whets and a few others also use this strategy when conditions are right.

Ordinarily females help feed their young from the time they are a few weeks old until they become independent, but males are the principal providers and can usually supply sufficient food if prey are readily available. Assuming her deserted mate can manage on his own, a female that engages in serial polyandry can significantly increase her reproductive output. One polyandrous long-eared owl produced twelve fledglings in a single breeding season: seven from her first clutch, which she started unusually early (on February 12), and five from her second clutch, initiated three months later. The average number of fledglings produced by monogamous females that nested in the same grove that year was only five.

Very rarely a non-breeding adult owl has been found helping a monogamous pair raise their brood by feeding the young and defending the nest site or territory. One case involved two male long-eared owls, both of whom were seen providing food to nestlings at the same nest. Through DNA analysis the biologists studying these birds proved that male 914 was the father of only two of the four nestlings, but they could not clearly identify the relationship between male 906 and the other two nestlings. All they could tell was that 906 was probably either their father or their half-brother. They suspected the latter scenario, hypothesizing that he was the son of an unknown male who had died or abandoned the nest after fertilizing the female's first two eggs. If so, male 906 was indeed a non-breeding helper and, like helpers in other bird species, he had a genetic stake in the fate of the nestlings he was helping to raise.

There is also one documented case of helping by a male eastern screech-owl. This owl was brought to a nature center in Connecticut after being struck by a car. Fully rehabilitated after a few weeks of care, he was then fitted with a radio transmitter backpack and released. Three months later, in mid-May, he relocated some distance from his usual haunts and began roosting close to the nest of two other eastern screech-owls, feeding their offspring. The non-breeding male seemed to make as many food deliveries as the presumed father, and neither member of the breeding pair showed any hostility toward the helper, even when food delivery visits by the two males briefly overlapped. After the young left the nest, all three adults continued feeding them, and twice they were seen perched together with the two fledglings. It is possible that the helper was related to one of the two breeding birds, but equally possible that they had no genetic links.

Breeding Patterns

Monogamous owls living in temperate latitudes normally nest only once a year. If the first nesting attempt of the season fails because of predation, poor weather or bad luck, owls of many species will not try again. When renesting attempts do occur, they are usually limited to situations where the first effort failed during incubation.

The relatively long breeding cycle of owls is probably the main deterrent to their renesting. If a pair misses the customary window of opportunity, often there simply isn't time for them to begin again within that breeding season. After an initial clutch is lost, the female may not be hormonally ready to ovulate again right away and she may lack the energy reserves needed to produce a second clutch, since female owls often lose a considerable amount of weight during incubation.

Double brooding—the raising of two consecutive broods by a single pair during one breeding season—is similarly constrained in temperate latitudes, but attempts are made now and then in North America, most often by barn owls and occasionally by burrowing owls in Florida and California. A few cases of double brooding by boreal and long-eared owls have been seen in Europe, but none are known from North America.

Barn owls in tropical regions frequently produce second broods, but this breeding strategy is uncommon in North America. In Utah, which is near the northern edge of this species' range, only about 5 percent of all pairs attempt a second nesting after their first young of the season have fledged. However, one remarkably productive pair in Illinois nested five times in two years and produced fledglings from all but the first of these attempts. Their first chicks hatched in December 1993 and died shortly thereafter during a cold snap. Further clutches were laid in March and August 1994 and February and July 1995. By the time the pair launched their last set of young, in October 1995, they had raised at least 15 offspring to the fledgling stage.

In contrast to the barn owl's ability to breed continuously under favorable circumstances, certain species, including spotted, snowy, great horned, northern hawk and boreal owls, may skip entire breeding seasons when conditions are poor, because of either prey scarcity or failure to secure a suitable nest site. The spotted owl may be one of the most irregular breeders in North America, with most pairs not breeding annually and some going up to six years in a row without nesting.

Prey population cycles clearly influence great horned owl breeding patterns in the boreal region, where the availability of snowshoe hares is the driving force, and in Utah's Great Basin desert, where black-tailed jackrabbit populations rise and fall dramatically. When these key prey species crash, few great horned owls nest. In other parts of North America, individual females forgo breeding about every third year. Snowy owls also depend on prey with a boom-and-bust population cycle, and some members of this species breed only every three to five years, when there is an explosion of lemmings.

The Pair Bond

For all owls, whether they mate for life or choose a new partner each year, courtship is a necessary prelude to nesting. In order to break down the normal social barriers between individuals and coordinate the team effort required for successful breeding, pair bonds must be established or renewed. Each species' breeding cycle is timed to coincide with peak prey availability during the nestling and fledgling periods, as well as to allow enough time for the young to reach independence before the next winter. In temperate latitudes, owls that feed primarily on mammals generally initiate egg laying in late winter or early spring, while insectivorous owls nest somewhat later. Courtship begins anywhere from a few weeks to several months before the ideal laying period, depending on the species.

Among the first to respond to their procreative urges each year are the great horned and great gray owls. Male great horned owls often begin regular territorial hooting as early as October, becoming increasingly vociferous as the nesting period approaches. The female's response also intensifies as the season progresses, and sooner or later the paired birds start roosting together near their chosen nest site, sometimes several months before the female starts laying. Male great gray owls may begin exhibiting courtship behavior as early as November, but pair bonds are usually established between January and April, and no later than two weeks before egg laying.

Spotted owls have a relatively long courtship period, commencing four to six weeks before egg laying. Mated birds usually maintain long-term pair bonds and share the same home range throughout the year, but they are mostly solitary during the non-breeding season. In February or early March they begin to roost together near their intended nest site and to call to each other at dusk and dawn as they set out and return from hunting.

The boreal owl's courtship period, which usually starts in January or February, may be delayed by prey scarcity or poor weather. Males breeding for the first time or newly arrived in an area tend to start singing later than experienced resident males. During the courtship period a male sings at or near up to five potential nest cavities until he has attracted a female and she has cemented the relationship by choosing one of his nest sites and moving into it prior to laying. The singing period for successful male boreal owls lasts from a few days to a couple of months. Those that fail to attract mates may continue their nightly broadcasts for more than three months.

Although migratory owls generally begin breeding activities somewhat later than sedentary owls, certain species have ways of making up time. Flammulated owls mate for life on the same territory; when one member of the pair dies, the survivor usually mates with a close neighbor that has also been widowed. This practice offers a level of familiarity that shortens and simplifies the pair-formation process. Males begin singing almost as soon as they reach the breeding grounds in late April or early May; when the females arrive shortly afterward, pair bonds are rapidly established.

Snowy owl courtship appears to be confined to the breeding grounds and to begin after the males establish territories in late April or May. However, a few observers have seen what they interpreted as incipient courtship displaying on the wintering grounds. Whether any pairs actually bond during winter and travel together to the Arctic remains debatable.

In addition to singing and calling, owl courtship behavior invariably includes presentations of prey by the male to the female. In some species, such as the snowy owl, courtship feeding is ritualized, incorporated into elaborate displays by the males. Female owls typically respond to the sight of a male bearing prey by giving food-solicitation calls and making body movements that are characteristic of juvenile begging behavior (vocal communication and displays are discussed in detail in chapter 4).

Initially courtship feeding may serve as a demonstration of the male's ability to provide for a mate and offspring. For many species it ultimately plays an essential role in building the female's energy reserves to prepare her for the demanding task of egg production. Once the pair bond is established and the male begins provisioning in earnest, his mate does little or no hunting for herself. With plentiful food and minimal activity, the females rapidly gain weight. Flammulated owls, for example, increase their mass by up to 68 percent prior to egg laying. At this point in the breeding season, females of this and other species often find flying difficult. In fact, burrowing owls can get so heavy they become grounded. Female cavity-nesting owls sometimes have to struggle to squeeze in and out of their nest-hole doorway.

Courtship feeding generally continues for at least a week, but often much longer. As the egg-laying period approaches, the female—particularly in cavity-nesting species—may take up virtually full-time residence in the nest, with her mate often roosting nearby during the day. This behavior is perhaps prompted by a scarcity of nest sites, but in cold climates it may also serve to warm the nest before the eggs are deposited. Boreal owls begin courtship feeding from one to three months before nesting, and the female occupies the nest cavity for a week to 19 days before laying. Female burrowing owls move into their nest burrows several days before laying. Being able to fly only ponderously or not at all, it is undoubtedly safer for them to stay out of sight.

Mutual preening, or allopreening, as it is known to ornithologists, is a significant part of courtship and pair-bonding behavior for many, if not all, owl species. Most observations of allopreening by owls have been of adult pairs. Allopreening also sometimes occurs between sibling nestlings and fledglings, especially in barn owls, and between parents and their young.

When allopreening, owls gently nibble at each other's head and facial plumage, sometimes sliding one or more feathers between their mandibles with a soft vibrating motion. These movements are often so rapid that the preener appears to be simply running its bill through its partner's feathers. The rubbing together of bills is also common. Both birds partially or entirely close their eyelids and nictitating membranes while allopreening, undoubtedly to avoid eye injuries.

Gently nibbling at each other's head and facial feathers, a pair of great gray owls enjoys a mutual preening session.

Preening pairs typically take turns offering and receiving these attentions, usually switching roles a number of times during a single alloproning session, which may last for several minutes. While being preened, the recipient often lowers or turns its head, as if to facilitate the process by exposing the nape or moving the crown or far side of the face to within reach of the preener's bill. One or both participants may call softly during the session. The pair's positioning during alloproning varies between species. Spotted owls typically perch side by side on a branch, facing the same direction. Great gray owls stand face to face on a branch or stump with breasts touching. Burrowing owls stand on the ground, facing each other.

In some species, such as the barn owl and western screech-owl, paired birds remain closely associated throughout the year and may preen one another at any time. Other owls alloproen during the breeding season only. Long-eared owls and ferruginous pygmy-owls regularly alloproen immediately before or after copulating. In barn owls this behavior is also common after copulation, but only by females to males. Spotted owls, in contrast, rarely engage in pre- or post-copulatory alloproning. Indeed, mutual preening is infrequent throughout the pair-formation, incubation and brooding periods. Only after the young have fledged in late May or early June and the pair has resumed roosting together do they regularly alloproen. For many pairs this is a daily activity until late September or early October, when males and females return to their largely solitary ways for the winter.

A few researchers have investigated the alloproning instincts of owls by participating in mutual preening sessions with their subjects. One captive female spotted owl, which imprinted on biologist Eric D. Forsman, would happily alloproen with him but would attack other humans that attempted such intimacy. When Forsman scratched her head, neck or face, she usually responded by "preening" his hand or lowering her head to present her neck or crown. If he didn't make the first move when she wanted to be preened, she would solicit his attentions by flying to perch near him, giving low calls and leaning toward him with her eyes partially closed.

Like her wild counterparts, this bird showed definite seasonal patterns in her receptiveness to alloproning. During winter she was only moderately interested and would often fly or hop away from Forsman when he scratched her, but by late February she was more responsive and began regularly initiating preening bouts. During July and August she tried to engage Forsman every time he entered her cage, and kept each preening session going for as long he was willing to take part.

When Robert Nero had wild great gray owls in hand during banding operations, he discovered that he could elicit mutual preening behavior by pushing the top of his head toward the bird's face. This gesture almost always prompted the owl to run its bill through his hair while gently nibbling his scalp and often

Old woodpecker holes or natural tree cavities provide shelter from the elements and protection from predators for cavity nesters like the northern hawk owl.

pulling a few hairs. In most cases the owl also responded by pulling back the feathers of its facial disk, raising the bristles around its bill, stretching its neck and then tilting its head to one side with its eyes partially closed—apparently inviting him to return the preening favor. Often the owl would bump its bill against Nero's head; he occasionally responded in kind, bumping his nose against the owl's bill, but was less inclined to do this after one owl bit his nose.

As the courtship period progresses, paired owls move into the sexual phase of the breeding season, but humans rarely catch them in the act. Not only do their largely nocturnal habits limit opportunities for observation, most owls' sexual encounters are fairly brief. Barn owl copulations last 10 to 20 seconds. Other owls take considerably less time to complete their union: from 4 to 7 seconds for great horned owls to 2 or 3 seconds for western screech-owls, long-eared owls and ferruginous pygmy-owls.

Just before and during copulation the female typically assumes a crouched or horizontal posture on a branch or on the ground, with her wings drooped and tail raised. While mounted on her back, the male usually flaps or flutters his wings, most likely for balance. A number of observers mention seeing the male use his bill to grasp at or "nuzzle" the female's nape feathers or scratch her head while mounted, but it is unclear whether these reports represent different interpretations of the same behavior. Both males and females close their eyes while copulating.

Pre-copulatory behavior varies between species. Prey presentation is an important part of the male's overture in some species—such as the barn owl, ferruginous pygmy-owl and possibly the short-eared owl—while others emphasize vocal exchanges, visual displays or mutual preening. The same individuals may vary their approach over the course of the breeding season or even during a single night. One flashlight-equipped observer watched a pair of western screech-owls during two copulation sequences on a January night and saw a marked difference in how each was initiated.

The first encounter began with both birds singing "bouncing-ball" songs from separate perches until the female moved in close beside her mate. For the next 10 minutes the pair sat side by side, singing frequently and allopreening. The male then suddenly switched to the species' secondary song and the two sang a brief duet that terminated when the male abruptly mounted the female. Their coition lasted no more than two seconds, with the male flapping continuously. The pair then separated by flying away in opposite directions. Half an hour later the male returned to the same branch and the female joined him. Forgoing any preliminaries, he mounted her immediately, this time appearing to try to grasp her neck with his bill. Moments later they again flew off in different directions.

Owls usually begin copulating well in advance of egg production and they copulate frequently during laying, ensuring that every egg in the clutch gets fertilized. In some species sexual activity persists long after the female has completed laying. Although copulation has no procreative function at this point, it helps reinforce the pair bond. Spotted owls start copulating as much as two to three weeks before the female begins laying, and they continue their sexual encounters until early in the incubation period. Barn owls copulate every few minutes when choosing their nest site, and with decreasing frequency throughout the breeding season. Ferruginous pygmy-owls also continue copulating after incubation.

Nest Sites

Selecting a nest site is an important part of the pair-formation process for owls. Males, and females of some species, scout out potential sites at the beginning of the breeding season or earlier—up to a year in advance of nesting, in the case of yearling barred owls that have not yet begun breeding. The selection process can be prolonged even once the choices have been narrowed down. Great gray owl pairs, for example, visit potential nests repeatedly during the two to four weeks before laying. In most, if not all, species the female makes the final decision on which site to occupy.

The great horned owl's habit of taking over stick nests made by other birds, such as great blue herons, occasionally makes for uneasy neighborhood relationships.

Most owls do no nest construction work, relying instead on prefabricated homes made by other birds or mammals, including humans, or naturally created sites that suit their purposes. This doesn't necessarily mean they aren't fussy about where they nest. Many species have distinct preferences for particular types of nest sites, though others are very flexible in their choices.

The most critical nest site requirements for cavity-nesters are suitably spacious interior dimensions and a doorway that is neither too large (and therefore vulnerable to invasion by predators) nor too small. Cavities carved out of trees or cacti by woodpeckers are generally superior to natural tree cavities created by decay and weathering, because they provide a more solid defense against enemies such as bears or raccoons that might try to break into the nest, and better insulation against cold and heat. Nevertheless, most cavity-nesting owls use natural cavities to some extent. Some also use nest boxes and various other artificial structures, from mailboxes to a pipe sticking out horizontally from tailings at an old mine site. A few will nest in holes in cliffs or earthen banks.

Northern saw-whet, boreal, northern hawk, flammulated and elf owls and pygmy-owls nest exclusively or primarily in abandoned woodpecker cavities.

The short-eared owl is one of only two true ground-nesting owl species in North America.

Screech-owls favor old woodpecker holes but also regularly nest in natural cavities. Large woodpeckers are the most important home-builders for owls because their nest holes are the right size. In Canada and the United States a large percentage of the cavities occupied by owls are the work of northern and gilded flickers and pileated woodpeckers. In northern Mexico the main cavity-suppliers are lineated and pale-billed woodpeckers and the two flicker species. Now missing from the picture are North America's largest cavity-creators, the almost certainly extinct imperial woodpecker and the extremely rare (or possibly extinct) ivory-billed woodpecker.

Small owls have additional accommodation options. The diminutive elf owl can comfortably inhabit holes made by woodpeckers as small as the ladder-backed and Arizona, and flammulated owls commonly nest in sapsucker cavities in some areas. Both of these owls also use cavities constructed by larger woodpeckers. Mottled, spotted and barred owls are too large to nest in woodpecker cavities and must settle for natural holes and hollows in trees. When these are unavailable, spotted and barred owls use open nest sites.

Barn owls nest in either natural cavities or alternatives provided by humans. These adaptable owls use a wide variety of natural nest sites, including naturally formed tree holes, openings in cliffs or rock outcrops and caves. In Mexico and the western United States they sometimes dig with their feet to excavate burrows in the soft soil of river banks or arroyos—a rare example of true nest construction by owls. The barn owl's long history of nesting in buildings is reflected in its common name in a number of languages besides English, including Dutch (*kerkuil,* or church owl) and Spanish (*lechuza de campanario,* or belfry owl). Among the oddest non-natural locations where these owls have nested are a drive-in movie screen, a large open chimney for a fireplace that was in constant use during the two seasons when owls nested there, an offshore duck blind in Chesapeake Bay, a wooden box sunk into the ground, and stone steps inside an old well—30 feet (9 m) below ground. Although barn owls usually choose fairly quiet, secluded spots to raise their families, one pair nested in New York's Yankee Stadium.

Long-eared, stygian, great horned and great gray owls principally use open, platform-style nest sites. They frequently take over old stick nests built by hawks, eagles, ospreys, herons, ravens, crows, magpies or other birds, and tree squirrel nests made of twigs and leaves. "Witches' brooms"—dense clumps of branches produced when parasitic dwarf mistletoe attacks live conifers—sometimes serve as nesting platforms for several open-nesting species, including long-eared, great gray and spotted owls. Other types of natural platforms used for nesting include cliff ledges and the concave tops of broken tree trunks.

Owls seldom use the same stick nest for more than a year or two because these structures usually deteriorate rapidly when not maintained and repaired. At best a particularly well-constructed nest in a firm tree crotch may last a decade before it disintegrates. Nest sites on cliffs, however, are often used for many breeding seasons. In Utah, certain cliff nest sites have been used by great horned owls for more than 25 consecutive years.

In addition to natural nest sites, great horned and great gray owls readily use purpose-built nest baskets or platforms. Great horned owls also sometimes nest in or on other human-made structures such as abandoned buildings and bridge beams. Possibly the most novel site ever used by this species was a flowerpot on the balcony of a third-floor apartment in a Kansas retirement home. The plastic pot, with a top diameter of 14 inches (36 cm) and depth of 12 inches (30 cm), was half-full of potting soil, so the female was easily visible as she sat on the nest incubating her two-egg clutch, which was laid in late February. By the time the two owlets fledged in late May the deck carpet was ruined, but replacing it probably seemed a small price to pay for this extraordinary opportunity to watch a pair of great horned owls raise their family.

Ground nesting is relatively uncommon among owls. In North America great horned, great gray, barred and long-eared owls occasionally nest on the ground, but the only true ground-nesters are snowy and short-eared owls. Unlike most other owls, these two species construct their own nests, a job that falls to the female. Snowy owls make their nests, known as scrapes, on turf or bare ground. The female first clears debris from the site by scratching with her claws, then swivels and twirls her body until she has formed a shallow but distinct depression. No lining of feathers or vegetation is added. Females sometimes inexplicably abandon a scrape after working on it for a few days, beginning again at a new site. Some scrapes are used repeatedly for many years. Short-eared owls, which typically nest in grasslands, employ similar techniques to create their bowl-like nests, which they line with grass or downy feathers.

The burrowing owl's nesting habits set this species apart from all other North American owls. The home builders that provide sites for these underground nesters include ground squirrels, badgers, prairie dogs, marmots, skunks, armadillos, kangaroo rats and tortoises. Burrowing owls are capable of excavating their own burrows, but they rarely do so, except in Florida. However, both males and females renovate and maintain their burrows, kicking the soil backward with their feet and sometimes using their beaks to dig. Burrowing owls often line their nest chambers with various materials layered as much as an inch (2.5 cm) deep. They also deposit these materials at the entrance hole and along the tunnel to the inner chamber. In some areas the most common material is dried cow or horse manure, which the owls shred. Feathers and grass may also be used, and one California population favored divots from a nearby golf course.

Nest preparations by other species are minimal, and supplementary materials are rarely added to premade nests. Barred owls may flatten or remove the top of an old squirrel nest before occupying it, or add lichen or fresh green conifer sprigs to a stick nest. Female barn owls often shred some of their own pellets and fashion a bowl to receive their eggs. Male boreal owls usually rearrange the material left by previous tenants in the bottom of their chosen nest cavity, shuffling it about and shaping a deep depression. This exercise, which they sometimes undertake a few weeks before egg laying or even before securing a mate, helps thaw, dry and warm the nest lining.

Owls often face stiff competition for nest sites, both from members of their own kind and from other birds or mammals. Woodpecker cavities and nest boxes in particular are highly desirable residences for a wide range of species, and disputes over ownership can be fierce. Elf owls have been known to displace ash-throated flycatchers and gila woodpeckers from nest cavities, but on other occasions have themselves been evicted from their chosen homes by acorn woodpeckers and western screech-owls.

The competitors most likely to usurp nests from eastern screech-owls are fox squirrels, European starlings and northern flickers. About a third of the time these owls reclaim the cavity later in the season, after the interlopers have finished raising their own families and departed. While the squirrels may eat the owls' eggs, starlings and flickers usually toss the eggs out of the cavity or cover them with new nesting material.

When owls take over active nests in cavities, they may have difficulty removing the previous occupants' eggs; since owls add no nesting material, the residual eggs remain uncovered. One flammulated owl was found incubating a pair of her own eggs in a cavity that also contained five bluebird eggs. Another laid her clutch alongside a northern flicker egg that perhaps belonged to the

Kicking vigorously, a Florida burrowing owl demonstrates its digging technique.

pair of flickers nesting in the same tree, 7 feet (2 m) above the owl's residence. The fate of the foreign eggs in these nests was not reported, but owls do occasionally hatch eggs that are not their own. In Georgia an eastern screech-owl incubated an American kestrel egg, which hatched at about the same time as her own three eggs. A pair of biologists examined the hatchlings when they were a few days old, at which point the kestrel chick appeared fine, but two days later it had disappeared. A mallard egg laid in a short-eared owl's nest that already contained three owl eggs was also incubated to hatching. The duckling was seen only on the day it emerged.

Competition for stick nests can also be severe. Great gray owls in the boreal forest appear to flourish in years when populations of other raptors are low, leaving plenty of vacant nests available. These owls generally lose out to great horned owls because the latter nest earlier in the year and have already claimed the best sites by the time the great grays are ready to settle down. But great gray owls are able to force out some competitors, such as goshawks, even after they have begun nesting. Long-eared owls are able to take over newly constructed crow nests but have been ousted by Cooper's hawks on at least one occasion. An extremely unusual stand-off between rivals for the same stick nest was discovered by ornithologist Charles Broley, who found a great horned owl and a bald eagle both incubating eggs on the same nest in Florida. Unfortunately a washed-out road prevented him from returning to the site, so there is no record of how this communal arrangement worked out.

CHAPTER SIX

Life's Journey

From the egg to independence, an owl's first months of life are full of change. Throughout this time both parents play vital and complementary roles in nurturing and protecting their brood. Once juvenile owls become self-sufficient they face many hazards and often perish before they are a year old. Those that survive their first year, however, may enjoy great longevity, their lives revolving around the annual cycle of raising their own offspring.

Eggs and Incubation

All owls' eggs are pure white, but they often become stained and discolored over the course of the incubation period. The eggs of cavity-nesting species are fairly round, while those of some platform-nesters and ground-nesters tend to be more oval, making them less likely to roll out of the nest. The great horned owl's almost perfectly spherical eggs are an exception to this general tendency. Owl eggs range in size from the elf owl's, which measure slightly more than an inch (2.5 cm) long, to those of great horned and snowy owls, which are about double this length. For many species the first egg in any clutch is the largest and the last is the smallest.

Clutches of two to four eggs are common for many North American owls, but several species have larger clutches on average or greater variation in clutch size. Among those with the most variability are short-eared and long-eared owls, whose clutch sizes range from one to eleven and from two to ten, respectively. For many North American owls there is an increase in clutch size from south to north through the species' range, though the differences may be small. In a few species clutch size increases slightly from east to west.

A flightless five-week-old great gray owl climbs to a safer perch.

Female elf owls lay one to five eggs, a significant physiological burden for these tiny owls.

Prey abundance significantly affects clutch size for a number of owls, especially those that depend on prey that regularly undergo dramatic population fluctuations. Great horned owls, for example, usually lay only one or two eggs but can lay up to five eggs when food is plentiful. Snowy owls respond strongly to food availability, producing clutches of three to five eggs when food is scarce and seven to eleven eggs during peak lemming years. Food availability accounts for some, but not all, variability in clutch size among eastern screech-owls, which lay from two to six eggs. Younger or smaller females typically produce smaller clutches than older or larger females. As well, when eastern screech-owls renest after their initial nest fails or is destroyed, their second clutch is generally smaller than the first, usually by one egg.

Egg laying often places a considerable physiological burden on female owls, since the weight of a single egg can represent a significant proportion of the mother's total body weight. In Texas a laying female ferruginous pygmy-owl will produce from three to seven eggs laid at intervals of 32 to 39 hours, with each egg weighing about 0.26 ounce (7.5 g)—10 percent of the average female's body mass. Some large owls are similarly taxed by their egg-laying efforts. A great gray owl's typical three-egg clutch corresponds to about 20 percent of her body weight. Female snowy owls can produce clutches that are equivalent to 43 percent of their own mass.

Compared to these owls and most other members of the Strigidae family, barn owls produce relatively small eggs for their body size. This economy means that a female barn owl can lay a clutch of seven eggs using same amount of body reserves that most strigid owls would use to lay five eggs. It also offers barn owls the flexibility to produce clutches of a dozen or more eggs when prey are abundant, or much smaller clutches, with minimal nutritional investment, when food supplies are limited.

Not surprisingly, female owls need to take a lengthy pause after laying each egg. The interval between eggs is rarely less than a day and is often two to three days or longer. Snowy owls usually lay their eggs two days apart, but they occasionally cease laying for up to five days before continuing at the regular pace; such gaps may be a response to inclement weather during the laying period. Laying intervals for great horned owls range from one to seven days, though two-day intervals are most common. In some species the interval gets longer as the clutch grows. Barred owls pause for 24 to 72 hours after laying their first egg, but for 48 to 96 hours after subsequent eggs. Eastern screech-owls lay daily for their first two or three eggs, then increase the length of their breaks until the clutch is complete.

Among owls, females take full responsibility for incubating the eggs and brooding the nestlings. In preparation for these duties each year, the female loses the feathers covering her brood patch, a specialized area of the abdomen and lower breast that is richly supplied with blood vessels. The bare skin of the brood patch becomes soft and cushiony to allow the greatest contact between this heat-delivery system and the eggs or, later, the nestlings.

Most North American owls begin incubating immediately upon laying their first egg. Exceptions include ferruginous pygmy-owls, which don't begin to incubate until the next-to-last egg is laid, and elf and northern saw-whet owls, which sometimes delay until the second egg. Among eastern screech-owls, 63 percent of females start incubating with the first egg, 25 percent wait until they have two eggs, and the remaining 12 percent start after laying three eggs. Burrowing owls in California and New Mexico commence incubating with the first egg, but in Saskatchewan and Oregon they may delay until the clutch is complete.

For owls that breed in cold regions or early in the year, an immediate start to incubation is essential. On the arctic tundra the weather is regularly severe enough during the incubation period that any snowy owl egg left unattended will freeze solid and crack open. But as long as the nest is in a windswept location where drifting snow will not accumulate, the incubating female will sit tight through cold and storms. The female snowy owl's brood patch is enormous and flabby, a necessity when trying to keep up to 11 eggs or owlets warm in such an inhospitable environment.

Great horned owls living at northern latitudes face similarly harsh weather during the early part of the breeding season, but incubating females remain at

Stoically enduring cold and snow, this incubating female great gray owl will hardly leave the nest until her offspring are a few weeks old.

their stations even when covered with ice or snow. They are able to protect their eggs through periods when temperatures drop to at least –30°F (–34°C). This species' eggs are also remarkably hardy; one female who was disturbed by an intruding male owl left her nest for 20 minutes when the air temperature was –13°F (–25°C), with no apparent ill effects on the eggs.

Ordinarily a female owl hardly leaves the nest from the time she begins incubating until her nestlings are a few weeks old. Her mate brings all of her meals, delivering them either directly to the nest or to a nearby feeding perch, depending on the species. Insectivorous owls cannot deliver large meals, so males have to make frequent feeding visits. The male elf owl supplies his sequestered mate with food just before her dusk recess and every two to five minutes throughout the night. A great gray owl, on the other hand, can satisfy his mate's hunger with three to five prey deliveries daily. Other than to receive food, females leave the nest only to cast pellets and defecate. Typically they take only one or two brief breaks in each 24-hour period, often at dusk and just before dawn.

The incubation period for most North American owls is about one month. Elf and flammulated owls have the shortest incubation time, ranging from 21 to 24 days. At the other end of the scale, great horned owls' eggs can take as long as 37 days to hatch, though the average period is 33 days.

In species that start incubating immediately, the eggs hatch sequentially and with approximately the same interval between hatchings as there was between laying dates. If the clutch is large, a wide range of ages and sizes of nestlings will share the same nest. In species that delay incubation until the clutch is complete or nearly so, the eggs hatch more or less simultaneously.

The length of time it takes owl chicks to enter the world is quite variable. Eastern screech-owls hatch in two to three hours. Barn owl chicks usually begin working to break free in the evening and are out by morning, sometimes with help from their mother, who breaks off bits of the eggshell. Snowy owl chicks hatch unassisted but slowly, taking up to three days from the time the first fine cracks appear in the shell until they emerge.

Female owls usually deal with cast-off eggshells by eating them or carrying them away from the nest site and discarding them. Sometimes, however, the shells remain in the nest and are gradually broken up by trampling.

Nestlings

Owls emerge from their shells wet, blind and helpless. With weak legs, a tiny, pot-bellied body and an oversized head weighed down by a stout bill, a newly hatched owlet cannot crawl or raise itself from the nest floor, though it can stabilize itself by extending its stubby wings. The bulbous protuberances that are its eyes will stay sealed for anywhere from a few days to more than a week, depending on the species.

Hatchlings appear almost naked at first, but their sodden feathers quickly dry and become fluffy. This initial plumage, which biologists refer to as the first natal down, is typically short and white or off-white and may be either dense or sparse. Within a week or two of hatching the first natal down is replaced by the second natal down—long, soft plumage that is often buffy or grayish and in some species is patterned with bars, mottling or other markings. The second down may darken as the chick gets older.

An owlet gains the ability to crawl feebly on its first day and can soon make the motions necessary to receive food: gaping, weakly taking hold of proffered morsels with its bill, and swallowing. But it will be one to two weeks before it can hold its head up strongly or stand erect. It also takes about that long before it can regulate its own body temperature. While nestlings are in this vulnerable state, the female continues her round-the-clock attendance at the nest, brooding her hatched young to keep them warm, feeding them, and protecting them from predators, as well as incubating any remaining eggs.

As long as the female is on the nest full-time, her mate must provide all the food for both her and their offspring. Males of some species feed older nestlings directly, but the female is always in charge of receiving and distributing food when the chicks are very young. Females prepare vertebrate prey by cutting and tearing them into small pieces and selecting only soft parts, including some internal organs, to feed the youngest of their brood. Initially they avoid feeding the nestlings any parts containing bones or fur. Mothers often crush invertebrate prey and remove appendages such as legs or stingers.

Photographers G. Ronald Austing and John B. Holt, who studied, banded and photographed great horned owls in Massachusetts, note in their book on this species how the females they observed were careful not to expose young nestlings to the cold at mealtimes. Instead of standing beside or facing the nestlings while feeding them, these mothers more or less maintained a brooding position: "She simply raises herself slightly over the young, draws the prey into the center of the nest, and carefully tears off small pieces which the owlets take by reaching out from under her breast feathers."

The growing nestlings quickly gain the ability to deal with indigestible prey parts by producing pellets. During their first few days, snowy owl chicks are fed only small scraps of flesh or other soft portions, such as heart, liver and testes, but by the time they are a week old some begin disgorging long pellets containing lemming fur and bones, indicating a change in their diet. At two weeks old snowy owls are capable of dissecting prey for themselves, but their mother still mostly handles this job, portioning out the food so that the older nestlings receive larger shares than their younger, smaller siblings. Two-week-old barn owls can swallow rodents whole, and they eat independently from this point on.

Owlets gain weight quickly, especially in their early days of life. Snowy owl chicks can increase their mass by 60 to 70 percent a day during their first three days, slowing by their eighth day to daily increases of about 20 percent and in their fourth week to 6 percent. In just a month a snowy owl chick can go from a hatchling weight of about 1.2 to 1.9 ounce (35–55 g) to a hefty 3 to 3.5 pounds (1.3–1.6 kg). Northern hawk owls gain about 0.3 ounce (9 g), or 13 percent of their own mass, every day during their first two and a half weeks. Other species also grow rapidly, though few match these rates.

Once nestlings can generate enough body heat to warm themselves, their mothers are no longer obligated to brood them 24 hours a day. At this point females typically begin hunting again, for both themselves and their young. With the appetites of the growing owlets increasing, this assistance relieves some of the pressure on the male, though he generally continues to provide a larger proportion of the food.

Owls that eat mainly invertebrates must make many provisioning visits to keep up with the nestlings' demands. Eastern screech-owls, for example, bring an average of two to five prey an hour. In southern areas their deliveries are most frequent around dusk and dawn; farther north, where summer nights are short,

the pace is fairly constant throughout the night. The average delivery rate for elf owls is once every two minutes, but the pace is even faster at dusk, when 12 to 20 minutes of intensive feeding is required to temporarily satiate the nestlings. For great horned owls a standard night's menu delivered to the nestlings might consist of five voles or a single hare or duck. Observed feeding rates for northern hawk owls range from two deliveries an hour at one nest in Manitoba to an average of about one delivery every two and a half hours at five nests in the Yukon.

The minimum period of continuous brooding for North American owls is about a week. Female whiskered screech-owls start leaving the nest to hunt when their young are about seven days old, northern pygmy-owls at about nine days old, and spotted owls at eight to ten days old. In contrast, female great gray owls are tied to the nest until their chicks are two to three weeks old, and ferruginous pygmy-owls spend a full three weeks brooding. Barn owls brood full-time until the first-hatched nestlings are about 25 days old. Although this is much longer than many other species, the youngest nestlings in a large brood may be only 11 to 13 days old at this time, and still in need of an external heat source. Females can safely leave these youngsters, however, because their older siblings will keep them warm. The same situation occurs with eastern screech-owls, which cease brooding when their oldest chicks are about two weeks old.

Just out of the nest, these fledgling flammulated owls are not yet ready to try their wings.

The fact that last-hatched eastern screech-owls may be kept alive by their siblings' body heat doesn't mean that family relationships are always benign. About a quarter of all eastern screech-owl broods experience some nestling mortality, and it is usually the youngest chick that dies, often while fighting its older, heavier siblings for food. Once dead, the deceased nestling is soon eaten by the others. Siblicide occurs most frequently in large broods and when there is a lack of cached food. The eastern screech-owl is not the only species in which nestlings die this way, but the full extent of siblicide among owls is not known, since it can be difficult to determine whether a chick was deliberately killed by a nestmate or simply became a convenient source of nourishment after succumbing to starvation or exposure.

The sacrifice of the youngest and weakest members of a brood in favor of the oldest and strongest is a natural response to unpredictable food supplies. Uneven-aged broods allow owls to hedge their bets in the annual reproductive sweepstakes, laying as many eggs as possible and, if all goes well, raising all of their young to maturity—or, if prey populations are very low, concentrating their efforts on ensuring the survival of at least some progeny.

Violent competition for food between nestmates is not inevitable. In fact, among northern saw-whet owls the reverse sometimes occurs in the latter part of the nestling period, when some mothers disappear from the scene. In these single-parent families, senior nestlings may dissect prey provided by the father and feed small pieces to their junior siblings. Similarly, older barn owl nestlings sometimes take prey brought to the nest by their parents and relay them to their smaller siblings, sometimes while uttering the adult food-offering call.

Between hatching and fledging, the nests of many owl species become fouled with droppings, pellets and prey remains. Elf owl nests may also contain the bodies of dead nestlings, which these insectivorous owls do not consume. Some owls, however, keep their nests quite clean for all or part of the nestling period, largely by the female eating the feces and sometimes the pellets of her offspring. Caching food away from the nest, rather than in it, also contributes to tidier abodes for species such as the northern hawk owl.

Female northern saw-whet owls practice meticulous nest sanitation while they are brooding continuously, but once the youngest nestling is about 18 days old the mother no longer roosts in the nest cavity, and conditions quickly deteriorate. By the time the owlets fledge two weeks later, the layer of pellets, feces and rotting excess food on the nest floor may be at least an inch (2.5 cm) deep.

In a number of species, including great horned, great gray and barn owls, young nestlings defecate in the nest and their wastes are eaten by the female, whereas older nestlings defecate over the edge of the nest. Barn owl nestlings also eject trampled and shredded pellets and other debris from the nest by scratching it backward out the doorway. Large amounts of whitewash-like excrement below a cavity entrance and piles of shredded pellets mixed with droppings on the ground make active barn owl nest sites easy to identify. In contrast, barred owls keep their nests and the area immediately surrounding the nest tree completely free of feces and pellets, offering no visual cues that might attract a predator (or interested humans) to the nest.

Out of the Nest

Owlets leave the nest at various stages of maturity, depending on the species. Some, such as long-eared and northern hawk owls, still wear their second natal down, while others are already clothed in their juvenal plumage. Juvenal body-contour feathers resemble those of adults, but are softer and looser. All strigid owls follow this progression from first and second down to juvenal plumage to adult plumage. In contrast, barn owls go directly from second down to adult plumage, completing this molt while still in the nest.

The flight feathers on the wings (the remiges) and tail (the rectrices) also begin growing during the nestling period, but there is considerable variation in how well developed they are upon fledging. When boreal owls launch from their nest cavity at 28 to 36 days old, their remiges are fully grown and their tail feathers are about 25 percent developed. They and the closely related northern saw-whet owl can fly reasonably well as soon as they leave the nest. Like most owls, they do not return to the nest once they have departed. When elf owls abandon the nest cavity 28 to 33 days after hatching, they can fly weakly, though well enough to catch crickets almost immediately. They are still fed by their parents for some time after their departure, but biologists are uncertain how long they remain dependent.

Barn owls make their inaugural flights when they are 50 to 55 days old. At first they fly clumsily and land awkwardly, but by 70 days their remiges and rectrices are fully grown and they have gained sufficient strength and agility on the wing to start capturing some of their own prey. Aerial competence is not the only skill needed for successful hunting, however, so the young owls spend a great deal of time pouncing on inanimate objects with a playfulness that disguises the seriousness of their practice. Fledgling barn owls remain closely associated with the nest for several weeks after their first flight, roosting there by day and receiving food from their parents at or near the nest until their 12th to 13th week of life, at which point they are cut off from the parental meal service.

Like barn owls, burrowing owls are gradual nest-leavers. At about two weeks of age they emerge from the nest burrow and begin waiting at the entrance for meals whenever their parents are away hunting. At three weeks old they begin running, hopping and flapping their wings, though they never stray more than a few feet from the burrow. A week later they become airborne for the first time. Their preliminary flights are short and rarely take them more than 50 yards (46 m) from home. By six weeks they can fly well. The fledglings initially practice hunting by jumping on dead and dying insects brought to the burrow by the adults, then graduate to chasing live insects at seven to eight weeks old. At this age they also begin to move between the nest burrow and nearby satellite burrows, which are used by both adults and young to escape predators and to shelter from inclement weather.

Other North American owls are more or less flightless when they leave the nest. Except for ground-nesting species, they depart by climbing onto the limbs of the nest tree or by falling, jumping or gliding from the nest. If they end up on the ground, they gain elevation by clambering up the trunks of leaning trees or seek refuge in low shrubs. Young flightless owlets—sometimes known as "branchers"—use parrot-like techniques to ascend trees and move about within the canopy, climbing foot-over-foot, grasping bark or twigs with their bill and vigorously flapping their wings. On the ground they walk or hop, sometimes using their wings as props.

The smaller owls are flightless for only a short time after they leave the nest. Flammulated owls fledge at about 23 days old, with the flight feathers about 60 percent of their full length on their wings and 50 percent on their tail; a day or two later they begin flying. Screech-owl nestlings move out into the world at about one month old. As with elf owls and a number of other species, the impetus to leave often comes from their parents, who call to them, withhold food or remove cached prey from the nest cavity. Eastern and whiskered screech-owls make their first flights two to three days after fledging, while western screech-owls reach this landmark in three to four days.

Like many owls, ferruginous pygmy-owls are dependent on their parents for food until they learn how to fly and hunt well enough to feed themselves.

Some ferruginous pygmy-owls leave the nest 21 to 24 days after hatching, but these early leavers have very poor chances of surviving more than a day. Fledglings that depart at 24 to 29 days begin making rudimentary flights four days later. During the first week out of the nest their flights are limited to about 11 yards (10 m) and conclude with clumsy landings. Flight distances double in the second week, and by week five they are flying freely and can capture lizards and insects.

Relatively little is known about nest departure by northern pygmy-owls, but it appears they can fly in a weak and uncoordinated manner when they leave the nest at about 23 days of age. At times, however, they may be seen hanging upside down from branches, a predicament that is common among inexperienced young owls of many species. Usually it results when the fledgling chooses a landing perch that cannot bear its weight, and is too confused or scared to release its grip when it finds itself in an inverted position.

Compared to the smaller owls, medium-sized and large owls tend to be flightless for longer after fledging, typically for about one to two weeks. Fledging age generally ranges from three to five weeks. The most prolonged period of flightlessness among North American owls is that of barred owls, which leave the nest when four to five weeks old and make their first short flights at ten weeks of age, progressing to more sustained flight by twelve weeks. Great horned owls have the longest nestling period—about six weeks—but are flightless for only a week after fledging. Some spotted owls leave their nests as few as 15 to 25 days after hatching. These premature leavers lack climbing skills and may spend up to 10 days on the ground. Typically, spotted owls fledge when 34 to 36 days old and can glide from tree to tree in steep terrain within three days. They achieve true flight at about 40 to 45 days.

Short-eared and snowy owls have very short nestling periods, possibly because ground nests are extremely vulnerable to mammal predators. Since these owls often produce large clutches, early departure of the older nestlings reduces crowding as the younger chicks grow. In both species owlets may leave the nest when as young as 14 days old, though snowy owls leaving at this age sometimes return to the nest one or more times before deserting for good. The latest departure for short-eared owls is at 17 or 18 days, while snowy owls may linger for up to 25 days.

Regardless of fledging age, these ground-nesting owls always leave on foot. Short-eared owls can fly by the time they are 28 to 35 days old. Young snowy owls move quickly and nimbly on the ground but do not master strong, controlled flight until they are at least 50 days old. During the learning phase, which begins at about 30 days of age, they make frequent short flights, especially down sloped terrain.

As soon as they abandon the nest, snowy owls scatter in all directions and take cover in depressions or among stones lying on the tundra. Since the female is still occupied with caring for the remaining nestlings, it is the father's responsibility to feed the fledglings. Although well concealed and spread out over 250 acres (1 square km) or more, they let him know exactly where they are hiding by issuing piercing cries whenever they are hungry. Once the entire brood has left home, both adults feed the fledglings for at least five weeks. As the young owls become airborne they converge on their parents any time they see them carrying prey.

Fledgling owls of most other species maintain much closer contact with their nestmates, at least initially, and some stay quite close to the nest for a while. Often they can be heard giving noisy food-begging calls at night. Ferruginous pygmy-owls are usually found within 11 yards (10 m) of their siblings, perched low in the tree canopy or in dense understory vegetation, for about five weeks after leaving the nest. Both parents feed the fledglings throughout this time, and the female usually stays within 55 yards (50 m) of the brood. The distance between the young owls increases once they become competent fliers, around week five.

Asynchronous hatching generally leads to asynchronous fledging, with nestlings leaving over a period of several days in the same order that they hatched. But western screech-owl broods, which hatch over a two- to four-day period, usually fledge in a single night. Parent western screech-owls remain close to their offspring for the next five weeks, feeding them and occasionally sheltering them with their wings during storms. Boreal owl parents accelerate the nest-leaving process by feeding their fledged young more frequently than those that remain in the nest. This neglect forces the last-hatched chicks to leave the nest two to three days earlier in life than the first-hatched. The entire brood then stays in a loose group within 110 yards (100 m) of the nest for a week or longer.

A young long-eared owl finds itself in a tricky situation after miscalculating the strength of a branch.

Owls that mainly eat invertebrates, like the flammulated owl, must make frequent feeding trips to meet the demands of their hungry nestlings.

In some species fledgling care is split between the parents. When only two young fledge from a flammulated owl nest, each parent takes charge of one fledgling. In broods of three or four, the owlets typically leave the nest in two groups over two consecutive nights; these factions then separate, each with a parent in attendance. The young flammulated owls become autonomous about 25 to 32 days after leaving the nest, but continue to hunt and roost with the siblings of their subgroup for up to 35 days after fledging. Whiskered screech-owls also divide their broods, but the two fledgling subgroups may be exchanged between the parents during the four or more weeks that the juveniles are dependent.

Rather than dividing broods, long-eared owls combine them. Flightless fledglings typically remain separate from each other until they become capable of limited flight, at about five weeks of age. After that they roost with their nestmates and sometimes also with fledglings from other broods. In one case the young from three different nests shared the same roost tree. Brood mingling may also occur among short-eared owls.

Female long-eared owls quit the nesting area when the fledglings are six to eight weeks old, leaving it to the father to continue feeding the youngsters until they are ten or eleven weeks old. Similarly, female great gray owls abandon their broods three to six weeks after fledging, and the males continue to feed and protect the young for up to three months. After a great gray owl relinquishes responsibility for her offspring, she often visits father-and-fledgling family groups in nearby territories, spending up to two days in their company. She may also occasionally drop in on her own family.

Independence and Beyond

As the hunting abilities of juvenile owls improve, parental feeding decreases, though fledglings may resist this push to independence. Young spotted owls, for instance, sometimes pursue their parents and beg in vain for food during August and September. At this point adult spotted owls seem to actively avoid their offspring, perhaps hastening their departure from their natal territory.

Juvenile dispersal is characteristic of all owls but is particularly evident among species that hold year-round territories, such as the spotted owl. Intensive study of spotted owls in the northwestern United States shows that juveniles rapidly move away from their parents' territory during the first few days or weeks of dispersal, then generally settle into temporary home ranges in late October or November. Between February and April many of the young owls studied made a second move, traveling considerable distances before settling again. For some spotted owls the area where they spend their second summer becomes their permanent home range, while others move on through a series of temporary homes before eventually establishing a lifelong territory when they are two to five years old. A spotted owl that lacks a permanent home may live as a non-breeding "floater" within an occupied territory until the same-sex member of a resident pair dies; the floater then takes its place and breeds with the surviving territory-holder.

In other species, such as the eastern screech-owl, dispersal is less complicated. As juveniles of this species gain independence, they start roosting farther from their parents and wandering beyond the boundaries of their parents' home range at night. Finally, at about two months old, they abandon their natal territory and go in search of a suitable overwintering area, usually 0.6 to 10 miles (1–17 km) away. Most juveniles settle within a week of leaving home. The next spring, some nest in or near their overwintering sites, while others move a short distance before settling down to breed.

In most bird species, including many owls, juvenile females tend to disperse farther than juvenile males, which reduces the likelihood that siblings will end up mating with one another. Female barn owls in Utah move 37 miles (61 km) on average from their natal site to their breeding territory, while the average dispersal distance for males is 22 miles (36 km). In the northwestern United States, average dispersal distances for spotted owls are 17 miles (29 km) for females and 12 miles (20 km) for males. And female juvenile eastern screech-owls travel about three times farther than males.

Juvenile burrowing owls sometimes disperse only a very short distance or not at all. In one nonmigratory Florida burrowing owl population, 36 percent of young males remained on their natal territory and bred there, but only 3 percent of the females did the same. The average distance traveled by the dispersing juveniles was 1,228 yards (1,116 m) for females and 455 yards (414 m) for males. In Alberta, where burrowing owls are migratory, some young birds return from their first trip south and find breeding territories as few as 330 yards (300 m) from their natal area, while others breed up to 18 miles (30 km) away. Typically the females disperse farther than the males.

For a number of owls the extent of juvenile dispersal is influenced by food supplies. For example, great horned owls in the boreal region disperse farther in years when snowshoe hares are scarce than in years when there are high populations of this key prey species.

Provided they survive their first winter, most North American owls breed as yearlings, but for several species sexual activity is commonly deferred. There are no records of barred owls nesting before they are two years old. Spotted owls, close relatives of barred owls, breed only occasionally in their first year, and even as two-year-olds they breed significantly less frequently than adults three years old or more. Great gray owls follow a similar course, breeding rarely as yearlings, occasionally at two years old and more commonly at three years old. Snowy owls are also thought to be delayed breeders, not mating until they are at least two years old.

Some great horned owls breed as yearlings, but a large proportion are unable to establish territories and must spend a year or longer as nonbreeders. These floaters spend most of their time along the boundaries of occupied territories, avoiding confrontations by remaining quiet and unobtrusive.

Death

An owl's first year of life is its most perilous, and many individuals die before they get their first opportunity to breed. Although nestling mortality is most often due to starvation, predators also take a toll on both eggs and young, despite the fierce protectiveness of parent owls, especially mothers. Cavity-nesters are usually safe from the depredations of hawks, crows and ravens, which often target open nests, but all owls are at risk of attack by tree-climbing snakes and mammals such as raccoons, squirrels, weasels and martens. Badgers are a major predator of burrowing owls, and both burrowing and short-eared owls often lose broods to domestic cats and dogs.

Nest parasites and other invertebrates can also be lethal. One northern saw-whet owl nest was severely infested with bloodsucking flies because it had been

After leaving the nest, juvenile snowy owls scatter across the tundra. Their piercing cries tell their parents where to find them.

built on top of a starling nest, and all four hatchlings perished. Mosquitoes and blackflies have been known to kill great gray owl nestlings.

Sometimes inclement weather is to blame for nest failure. A heavy downpour apparently flooded two whiskered screech-owl chicks from their nest; the pair died from either drowning or hypothermia after a day on the ground. In Florida, spring rains often cause sudden collapse of burrowing owl burrows, trapping adults and owlets underground with no means of escape.

Once they leave the nest, young owls face the same dangers as adults, but they are more vulnerable than their elders because they are poor fliers and unskilled at hunting, and lack experience in detecting and evading predators. Biologists who studied burrowing owls in North Dakota found that two-thirds of the juveniles that died in their first year of life succumbed before the autumn migration, with mortality rates peaking in two distinct periods. Swainson's hawks, badgers and other predators were responsible for most of the deaths in the first period, which occurred when the five- to seven-week-old owls were learning to fly. No fledglings died over the next ten days, but fatalities increased again during the following two weeks. Starvation appeared to be the main cause of death during this period, when the young owls were fully responsible for feeding themselves for the first time.

Overall, only about 20 to 30 percent of all burrowing owls survive their first year; annual survival rates for adults are much higher. This pattern is typical of most owl species whose demographics have been studied. Typically an individual that makes it to one year of age has a relatively high probability of living for many more years. In North America, barn owls are the one exception to this rule. Up to 75 percent of all barn owls die in their first year, before they reach breeding age, and mortality rates remain quite high in the following years. Banding data from across the continent show an average age at death of 21 months, and in a 20-year study of barn owls in Utah, 85 percent of the adults survived only one breeding season. The maximum documented lifespan in the Utah population was eight years.

Aside from one 34-year-old barn owl, which was clearly an extraordinary representative of its kind, the most long-lived North American owl is the great horned. Two great horned owls that were banded as nestlings lived to just over 21 years and 22 years, and another that was banded as an unknown-age adult died 28 years and 7 months later. Survivorship for this species varies widely between years, depending on food supply, and territory-holders fare much better than floaters.

Other long-lived species include barred, spotted, long-eared and snowy owls. The record for barred owls is 18 years, while spotted owls can live at least 17 years in the wild and 25 years in captivity. One captive snowy owl lived to be 28 years old, though the longevity record in the wild is just short of 11 years. For long-eared owls the greatest known lifespan is 27 years and 9 months in Europe and just over 11 years in North America. Smaller owls tend to have shorter life expectancies, but even elf owls, whose maximum recorded lifespan in the wild is only 4 years and 11 months, can live up to 14 years in captivity.

The most common natural causes of death for adult owls are predation and starvation. Fluctuating prey populations often lead to the latter, but poor weather can also play a role. Flammulated owls, for example, migrate south to ensure access to insect prey during winter, but they occasionally die during unseasonable spring snowstorms that cut off their food supply. Barn owls frequently die during severe winters, when deep snow makes it difficult or impossible to catch prey and cold temperatures increase their need for food. Less commonly, weather conditions alone can kill, as happened when a severe hailstorm in Saskatchewan took the life of four burrowing owls.

Owls fall prey mostly to birds or mammals and occasionally to snakes. As nocturnal hunters, owls are among the main predators of other owls. Great horned owls are themselves safe from most predators once they reach adulthood, but are one of the main killers of great gray, barred and spotted owls. Smaller owls must be on guard against attack by both medium-sized and large owls. Hawks often target owls that are active during daylight hours; northern goshawks and red-tailed hawks can kill owls as large as the great gray. The main avian predators of snowy owls are jaegers.

Except for incubating and brooding females, adult owls generally have little to fear from mammals. Significant four-legged predators of nesting females include foxes for short-eared and snowy owls, badgers for burrowing owls and martens for boreal owls. Great gray owls are occasionally killed by lynx, most likely when the owl's own hunting efforts have brought it to the ground. One unlucky northern pygmy-owl met its end when it attacked a weasel, which turned on the owl and killed it.

Among the few reports of an adult owl being killed by a snake is a great horned owl that was discovered wrapped in the lethal embrace of a dead southern black racer in Arkansas. The biologists who found the pair concluded that the owl had attacked the snake and inflicted wounds that were serious enough to kill it, but not before the snake managed to strangle its adversary. In Arizona a gopher snake was spotted just after it had seized an unsuspecting western screech-owl from its roost in an oak tree. The snake hung from a branch with its body coiled around the struggling owl for at least four minutes, until the owl grew still. The observers then intervened and shot the snake. When it fell to the ground they were surprised to see the owl free itself and fly away, seemingly uninjured.

Shooting owls was once legal and common, but this practice is now unlawful and comparatively rare. Unfortunately, other types of human-related mortality, particularly vehicle collisions, remain significant for many owl species. Because burrowing owls regularly sit on and hunt along roads at night, they are frequently hit by cars and trucks. Motor vehicles were to blame for a quarter of all deaths in one Florida population and 37 percent in a Saskatchewan study. Automobile collisions are also a major cause of death for barn owls in many parts of the world. In Great Britain, vehicular mortality has completely eliminated barn owl populations from areas within 1.5 miles (2.5 km) of major roads. The western screech-owl's susceptibility to being hit by vehicles may relate to its propensity for hunting for earthworms on wet roads, at least in the Pacific Northwest. Fatal accidents involving other forms of transportation are much rarer. Northern hawk owls are occasionally hit by trains, but a bigger problem is the collisions that regularly occur around airports between planes and both short-eared and snowy owls.

Other relatively common human-linked causes of owl mortality include colliding with transmission lines (with death resulting from either impact injuries or electrocution), becoming entangled in barbed-wire fences, and getting caught in leghold traps (while scavenging on carrion placed in the trap as bait). Deaths from secondary poisoning as a result of eating contaminated prey—such as mice or rats that have been exposed to anticoagulant rodenticides or pigeons that have eaten strychnine-laced grain—are rare, but do occur. Toxic substances can also cause behavioral changes that may ultimately lead to death or may reduce reproductive success. Concerns about this type of indirect effect have prompted Agriculture Canada to prohibit farmers from spraying the insecticide Carbofuran within 275 yards (250 m) of occupied burrowing owl nest burrows.

CHAPTER SEVEN

At Rest and on the Move

We often encounter owls when they are asleep or resting during the day. Whether you are alerted to the presence of a roosting owl by a mobbing flock of birds, by telltale accumulations of pellets and white-wash, or by chance, the discovery frequently provides an excellent opportunity to study one of these elusive birds and perhaps to watch it groom itself.

Owls in motion offer a different kind of interest, even when seen fleetingly or obscured by darkness. The buoyant, moth-like flight of the foraging barn owl; the deep, powerful wingbeats of the great horned owl as it lifts off; the elf owl's high-speed flutter—the way an owl moves is as much a part of its personality as its voice or eye color.

Biologists are also interested in owl movement on a larger scale, but they have yet to fully penetrate the mysteries of migration and nomadism for most species. Where do members of certain breeding populations spend the winter? What routes do migrants follow? What triggers peripatetic wandering, and which members of a population are most likely to roam? The questions are many and the answers slow to emerge, even for a species such as the northern saw-whet owl, which has been the subject of intensive autumn migration banding programs for several decades.

Roosting

When birds sleep or rest, they are said to be roosting. In general, nocturnally active owls roost during the day and diurnally active owls roost at night, but owls also take rest breaks during their active periods.

With powerful wingbeats, a snowy owl comes in for a landing.

Nocturnal owls typically leave their roosts around sunset, though they may depart earlier if driven by hunger or the demands of feeding nestlings, or when heavy cloud cover darkens the sky prematurely.

Most owl species have characteristic roost sites, which are generally selected to provide two things: concealment from enemies and shelter from the elements. Many owls roost on tree branches, choosing sites that are hidden by foliage. Usually these are close to the trunk, where the owl's cryptic plumage blends with the tree bark, but northern saw-whet owls frequently perch near the end of a sturdy limb in a spot that is veiled by overhanging foliage. Shrubby thickets and vine tangles are also commonly used for roosting by a number of species.

Compared to branch roosts, cavities provide better protection against wind, rain and extreme temperatures, but poorer security against attacks by tree-climbing mammals or snakes. Adult males of some cavity-nesting species, such as the flammulated owl, never roost in cavities, and their mates do so only while incubating and brooding. Males of some other species limit their cavity roosting to inclement weather. Eastern and western screech-owls typically roost in deciduous trees in summer and in conifers or cavities in winter.

One study of barn owls showed that the metabolic demands of heat production decreased significantly when the owls used sheltered roosts rather than remaining out in the open, with a reduction of 10 percent when roosting in a spruce tree and 19 percent when roosting in a building—in this case, an empty stable. The owls' energy savings were greatest when the weather was wet and windy and smallest when it was dry and calm. At night barn owls often return to their roosts between foraging bouts, allowing them to stay warm at a lower energy cost than if they stayed outside.

Spotted owls are fairly susceptible to heat stress, so in summer they usually roost in cool, shady areas of the forest, often near streams. Their summer roosts are typically in the lower part of the forest canopy. In winter they choose sites that are higher and closer to the tree trunk, and they sometimes bask in shafts of sunlight. Rather than staying in one place throughout the day, spotted owls often shift short distances from one roost to another to find more favorable conditions as the air temperature or sunlight patterns change. The search for cooler temperatures probably influences the mottled owl's roost site selection as well. These owls, which roost on tree branches or among vines, often spend the day perched within 6.5 feet (2 m) of the forest floor, especially during very hot weather. Western screech-owls that live in hot regions with few trees sometimes retreat to caves or rock-crevice roosts.

Although owls generally shun exposed roosts, such sites are sometimes advantageous. Great gray owls seeking warmth in winter may settle atop leafless trees in open, sunny areas. In summer, great horned owls trying to avoid blackflies sometimes roost on breezy open perches.

To avoid heat stress in summer, spotted owls roost in cool, shady parts of the forest.

Burrowing owls roost in a burrow entranceway or a depression in the ground. Because they nest in largely treeless habitats, short-eared and snowy owls usually roost on the ground during the breeding season, selecting slightly elevated sites when they can; in winter, short-eared owls sometimes roost in trees. The ability of snowy owls to sit motionless for hours on roosts that offer no protection from wind or cold temperatures is remarkable. Their insulation is equivalent to that of the Arctic fox, one of the best-insulated northern mammals. The only bird known to have lower thermal conductance (a measure of how much body heat is transmitted to the air) than the snowy owl is the Adélie penguin.

Whether an individual owl uses the same roost repeatedly or switches sites from day to day depends on species tendencies and, in some cases, the time of year. Outside of the breeding season great horned owls show little faithfulness to their roost sites, typically finding a convenient roost wherever their foraging ends at dawn. During nesting, however, breeding males usually have a preferred roosting site close to the nest and may use the same site year after year. Northern saw-whet owls take the opposite approach: during fall and winter these owls often use the same roost on a daily basis for weeks or months at a time, whereas during the breeding season mated males mostly use a different roost each day.

The loyalty of nonbreeding northern saw-whet owls to favorite roosts is so strong that they often return to them day after day. Certain roosts seem to hold a special appeal to more than just a particular individual. In Massachusetts, Ronald Austing and John Holt counted seven saw-whets that roosted in the same 10-foot (3 m) Scotch pine on different occasions over a three-year period, and each bird chose exactly the same spot on the same branch. Favored roost sites of these and other owls can easily be identified by the large concentrations of pellets and whitewash (droppings) that build up below them.

Boreal owls rarely use the same daytime roost twice. Like great horned owls, they appear to roost wherever they end up at the end of their night's hunting. In Idaho, biologists studying this species found that the average distance between roosts used on consecutive days was about a mile (1.6 km) in winter and slightly more than half a mile (800 m) in summer. During the breeding season males usually roosted more than 0.6 mile (1 km) from their nests.

Male western screech-owls roost relatively close to their nests from the start of the laying period. After the eggs hatch they find sites that are even closer, often within 16 feet (5 m) of the nest. Once the female stops brooding, both parents roost opposite the cavity entrance, where they are well positioned to defend their young against predators.

For the most part adult owls roost alone, but recently fledged young often roost with their siblings, perched side by side or in close proximity to each other. The length of time that this association lasts varies between species. Boreal owls, for example, roost in the same tree as their nestmates for only three weeks after fledging. Juvenile great horned owls roost together for two months after leaving the nest.

In many species it is also common for one or both parents to roost near their offspring until they become independent. Mottled owls are among those that roost in tight family groups after the young fledge; at this time of year the two parents and all of their fledglings are often found roosting within 3 feet (1 m) of each other.

When adults roost in the company of others, it is usually with their mates. At various times during the breeding season, mated pairs of a number of species—including spotted, great horned, mottled, burrowing, elf and great gray owls—may roost side by side. In winter, mated eastern screech-owls occasionally share a roosting cavity. Long-eared owls are unusual in that breeding males with neighboring nests occasionally roost together during the nesting period.

During the nonbreeding season, long-eared and short-eared owls often roost communally with other members of their own species, and sometimes the two types of owls occupy the same roost tree. The distance between individuals in communal roosts is often less than 3 feet (1 m). While short-eared owls sometimes roost in trees in winter, they may also roost on the ground, either alone or communally. They usually choose sites in dense grass that is less than 16 inches (40 cm) high, often in the shelter of a grassy tussock.

Long-eared owls roost only in trees, usually perching between 1.6 and 16 feet (0.5–5 m) above the ground. Typical long-eared owl communal roosts contain 2 to 20 individuals, but as many as 80 to 100 owls have been found roosting together. Although long-eared owls often show up at the same communal roost sites year after year, there is a high turnover of individuals from one winter to the next.

In contrast to communal roosting, some owls occupy cavity roosts on a sort of time-share basis, though the arrangement is purely one of ad hoc convenience rather than planned cooperation. While conducting an ornithological survey in southern Arizona and northern Mexico, Joe T. Marshall Jr. observed that elf owls inhabiting the pine/oak woodlands regularly roosted in cavities made and occupied by acorn woodpeckers. "Consequently," he wrote, "there is a great din and mutual protest at dusk when the owls emerge and the woodpeckers retire in the same trees."

Owls generally sleep lightly, remaining alert and regularly opening their eyes to scan their surroundings. Close observation of roosting boreal owls by one researcher showed that they spend their days perched quietly but rarely sleep for more than 40 minutes at a time. During their two- to five-minute breaks between naps they mostly look around or preen.

When sleeping, an owl typically draws its head down toward its shoulders, fluffs up its body plumage and relaxes or "folds" its facial disk. Owls with ear tufts lower them. When disturbed while roosting, a number of owls—including the northern hawk, great gray, short-eared and long-eared—may adopt an erect

Squinting and standing tall, an eastern screech-owl responds to a disturbance by making itself as inconspicuous as possible.

posture that is very different from their sleeping or relaxed stance. The owl raises its ear tufts (if it has them), elongates its facial disk and narrows its body profile by standing tall, compressing its contour feathers and holding its wings in tight. It may also erect its white eyebrow feathers and the bristles around its bill.

When northern saw-whet, boreal and elf owls and northern pygmy-owls assume this upright pose, they sometimes raise one wing to cover the lower half of the face like a Hollywood Dracula and peer over the top with wide-open eyes. Most other owls also keep their eyes open, but screech-owls squint. With its eyes reduced to mere slits, a screech-owl perched close to a tree trunk in an erect posture is amazingly well camouflaged.

Mobbing

Despite trying to be inconspicuous, roosting owls are sometimes discovered and mobbed by birds that do not take kindly to their presence. Often more than one species participate in these raucous group demonstrations of displeasure, diving at or loitering around the perched owl while calling noisily. Biologists suspect that mobbers are motivated by a desire to drive off a potential predator, though mobbing may also have an educational function, teaching less experienced birds to identify their enemies. Mobbing certainly serves to inform other birds—and humans—that there is a roosting owl in the neighborhood. Mobbers also sometimes direct their attentions to owl nests or nest cavities.

Not all owls are equally susceptible to being mobbed. Great horned owls elicit intense mobbing from a wide variety of birds and are particularly harassed by crows, which will zero in on any great horned owl they spot and often pursue it from tree to tree, cawing loudly. A mobbed great horned owl usually flies only a short distance from its original perch, moving to the closest site that offers better security. Most small birds keep their distance when pestering owls, but some, such as flickers, jays and kingbirds, may fly directly above or behind the fleeing owl and occasionally close in to peck at it. If a great horned owl that is being chased by crows, ravens, hawks or falcons lands on the ground or an exposed branch or ledge, the mobbers may swoop down at it, prompting the owl to respond with a threat display.

Although similar in appearance to great horned owls, long-eared owls are seldom mobbed, perhaps because they do not commonly prey on birds. However, diet does not necessarily determine which species are targeted. Small birds, ranging in size from hummingbirds to robins, sometimes mob whiskered screech-owls, elf owls and flammulated owls, all of which are primarily insectivorous and pose no threat to them. These owls may be the victims of mistaken identity, since eastern and western screech-owls do include birds in their diet.

Northern pygmy-owls are mobbed fairly frequently and generally ignore this provocation. They are unlikely to suffer any harm from small tormentors such as hummingbirds, wrens or warblers, but one northern pygmy-owl was apparently killed by a gang of eight to ten gray jays and a Steller's jay, and the same group of birds chased a second pygmy-owl into the forest.

Although owls are usually on the receiving end of mobbing behavior, they sometimes take the offensive role. The long list of species that short-eared owls have been seen mobbing includes snowy owls, bald eagles, gulls, black ducks, American bitterns, great blue herons, turkey vultures, red-tailed and rough-legged hawks, American kestrels, badgers and red foxes. Elf owls mob gopher snakes, great horned owls and ringtails when these potential predators venture too close to their nests. These tiny owls mob cooperatively, with up to four neighbors joining the nesting pair, all of them uttering barking calls and aiming attacks at the intruder's head.

Burrowing owls may also join forces with their neighbors to launch mob attacks on perceived enemies. One researcher working in New Mexico found that breeding adults usually ignored predators outside their own territory but would come from as much as 330 yards (300 m) away to mob a great horned owl. He also noted that male burrowing owls tolerated their associates during mobbing bouts instead of performing the customary territorial displays. In California another observer watched a multi-family group of 10 adult and juvenile burrowing owls chase a weasel across a pasture, hovering above it as it ran and swooping down with claws extended, though it was unclear whether they actually struck the weasel.

Comfort and Maintenance

Owls, like most birds, perform a variety of activities that are collectively referred to as comfort and maintenance behaviors. These include preening, scratching, stretching, bathing, sunning and dusting.

When an owl preens, it sweeps its bill through its plumage, often grasping individual feathers and drawing them through its mandibles. These actions serve to remove dirt, debris and external parasites from the feathers and skin. As it manipulates the feathers the owl also works in oily secretions produced by the uropygial gland, which is located on the rump. Preening keeps plumage in prime condition and ensures that all of the feathers are properly arranged. Near the end of a preening session owls often give a few vigorous shakes to adjust their plumage.

Preening is typically accompanied by scratching: the owl uses its toes to clean its head and face, often paying particular attention to the areas around the bill, which may be soiled with prey remains, and around the ears. Barn owls use the serrated edge of their middle claw like a comb when scratching. Owls, especially carnivorous species, also need to clean their feet regularly. They do this by using the bill to pick and scrape material off the toes, talons and foot pads.

Owls preen and scratch periodically during periods of wakefulness while roosting, and usually conduct a thorough grooming session before leaving the roost in the evening (or morning, in the case of diurnal species). Great horned owls usually preen after a flight and sometimes before flying. They also preen more frequently in windy weather.

Stretching is often an adjunct to preening. Like athletes warming up before a game, owls usually stretch their wings and legs before setting off to hunt and during breaks between foraging bouts. Standard exercises include extending the wing and leg on one side of the body while standing on the opposite leg, and stretching both wings over the head or trunk while standing on both legs.

For many owl species, bathing in the wild remains undocumented, but all owls probably clean themselves this way, at least now and then. In captivity owls often bathe daily. Although most owls bathe alone, eastern screech-owls often perform their ablutions with several others of their kind.

Owls normally bathe in shallow streams or at the edge of a lake or pond. Small owls have also been known to use backyard birdbaths, water barrels (in which they sometimes drown) and livestock watering troughs. Regardless of the venue, the method of bathing is consistent: the owl repeatedly dips its head and breast into the water and flips water over its back; it may also splash the water around with its wings and feet. After emerging it ruffles and shakes its feathers to shed any remaining droplets. Burrowing owls, which typically have little access to standing

Like an athlete warming up before a work out, this short-eared owl stretches its legs and wings in preparation for hunting.

water, become very excited during rain showers. Adults and fledglings alike run around stretching and flapping their wings, then stop to shake and preen.

But water is not essential for maintaining plumage in optimal condition. Burrowing owls take regular dust-baths in loose dirt or sand, as do short-eared owls, ferruginous pygmy-owls and probably other owls that inhabit arid areas. A dust-bathing owl typically stands or squats on the ground, then pushes its head face forward into the dirt, using the leading edge of its wings for balance. With rapid head movements it flips soil particles over its back. This is followed by energetic feather fluffing, which works the soil through the plumage. Short-eared owls take dust-baths both communally and on their own.

In winter some owls snow-bathe. Tom J. Cade described this activity in detail after watching a northern hawk owl one March morning. The owl landed on top of a telephone pole capped with 2 inches (5 cm) of fresh dry snow and settled down with its wings partially spread, tail raised and body feathers fluffed out. It then "commenced a vigorous bathing. With its face partially buried in the snow, the whole head and body were shaken, thus scattering the snow generally all about, much of it falling into the air. But certain specific movements of the head were directed to throwing bits of the snow over its back and wings." The owl carried on this way for two or three minutes, pausing intermittently to stand erect for a few moments, then continuing its bath.

It seems unlikely that birds that try to hide away by day would enjoy sunning themselves, but many owls do, especially during cool weather or after heavy rains. When sunbathing, owls orient themselves to the sun's rays and close their eyes, or sometimes they adopt a particular posture that maximizes their exposure. Short-eared owls face the sun with head extended forward, eyes closed and wings spread. In winter, eastern screech-owls often sit in a roost-cavity entrance facing the sun with their breast plumage fluffed up and facial disk feathers erected. One of the very few observations of a barn owl sunning itself was of a bird that lay on the ground in a small patch of sunlight and raised one wing to fully expose that side of its body to the rays.

Locomotion

Compared to other birds, most owls have large wings relative to their body mass or, in ornithological terms, low wing-loading. This trait affords several notable advantages. One is increased flying agility, an aptitude that is especially important for forest-dwelling owls. Another is reduced aerodynamic noise, an obvious benefit when hunting vertebrates in the dark, but apparently superfluous for insectivorous owls. Low wing-loading also facilitates prey transport, permitting owls to carry relatively heavy prey without being

overburdened and to retain their maneuverability while carrying somewhat lighter kills.

Owls also have comparatively short, broad wings. This characteristic is most pronounced in species that hunt among dense vegetation and need to be able to make quick turns and dodge between branches, such as barred, spotted and northern saw-whet owls. Owls that hunt by coursing over open country, such as short-eared, long-eared and barn owls, have longer, less rounded wings.

Certain owl species have distinctive flight styles that can be helpful as aids to identification. Pygmy-owls and northern saw-whet owls alternate between rapid wingbeats and brief closed-wing glides, resulting in an undulating, wood-pecker-like flight pattern. Screech-owls fly with uniformly rapid wingbeats, rarely interrupted by gliding. When hunting, short-eared owls mostly cruise in low quartering flights over open ground with their wings held at a slight angle above the back, now and then interjecting a few slow, deliberate wing-beats. Long-eared owls also course back and forth across open ground when hunting, but their wingbeats are faster and they usually hold their wings level while gliding.

A number of species, including screech-owls, flammulated owls, ferruginous pygmy-owls and northern hawk owls, have an idiosyncratic way of flying away

Night-time travel makes it difficult to study migration of species such as the long-eared owl.

from nest cavities and roost sites. The owl dives steeply from its departure point, flies low to the ground for some distance, and then rises abruptly to another perch. This U-shaped flight path, with its sudden beginning and end, helps obscure the location of nests and habitual perches.

Although they are adept fliers, owls sometimes employ other methods of locomotion, especially hopping, walking and running. When owls walk, they lurch along with a pronounced rolling gait. They run with fully extended legs, sometimes with the assistance of extended or flapping wings.

Adult screech-owls and elf owls sometimes clamber around in trees with the same parrot-like climbing techniques used by fledglings of these and other species. All cavity-nesting owls undoubtedly use their feet and beak to scale the inside walls of cavities. Barn owls are very good climbers and can make long ascents up the vertical internal walls of silos and hollow trees.

Owls are not swimmers by choice, but they sometimes find themselves swimming by necessity. One large but still flightless juvenile snowy owl escaped a pursuer by leaping into a lake and using its wings to paddle 50 feet (15 m) to the opposite shore. A fledgling long-eared owl that accidentally put down in a river was fairly feeble in its swimming efforts, but it did get to dry land through a combination of wing-flapping and floating. Adults can also swim if they have to, as demonstrated by a male short-eared owl that was so distracted by its own distraction display that it crash-landed into a shallow pond. It used its wings to row the 33 feet (10 m) to shore, then stood on the edge of the pond and continued its distraction display and calling.

Migration and Other Long-Distance Travel

Migration is the exception rather than the rule among owls. Worldwide only about 20 species have been identified as true migrants, meaning that they leave the breeding grounds after the breeding season and move to a separate wintering area, then return to the same nesting area the following year. Furthermore, none of these species is completely migratory, which would involve vacating the entire breeding range during the nonbreeding season. In most cases a portion of the species' range contains populations that are resident year-round. For some species, such as the snowy owl, the extent of migration depends on local conditions; a few individuals may remain on the breeding grounds during favorable winters.

Overall a majority of migratory owls depend primarily or exclusively on invertebrate prey; in North America this includes burrowing, elf and flammulated owls. Among the non-insectivorous migrants are northern saw-whet, short-eared and snowy owls. Long-eared and barn owls that breed in some northern regions of the continent are also migratory.

Instead of making north/south (or, less commonly, east/west) journeys, some owls migrate vertically, moving between high-elevation breeding grounds and lower wintering areas. These elevational migrants include certain populations of the northern pygmy-owl, whiskered screech-owl and great gray owl, as well as some California and Mexican spotted owls.

Very little is known about the migratory behavior of most North American owls, since they mostly move under cover of darkness. Nighttime mist-netting and banding operations have supplied most of the data collected so far. However, at the Hawk Ridge banding station in Duluth, Minnesota, migrating long-eared owls are often seen just after sunset, flying 100 to 165 feet (30–50 m) above the ground. A visual study of migrating owls conducted at Cape May Point State Park in New Jersey relied on observations made with the naked eye or binoculars by moonlight, sky glow from the nearby city, streetlights and the revolving beam of a lighthouse, as well as with a night-vision scope. One evening during the Cape May study, 17 long-eared owls took off from various locations in the park shortly after sunset and coalesced into a loose flock before heading out over the bay. Barn owls were also seen flying in loose flocks past Cape May; they called frequently and loudly as they flew and were often answered by other barn owls, perhaps coordinating their movements. Elsewhere, small flocks of migrating elf owls have been seen on several occasions.

Another type of seasonal movement common among owls is sporadic wandering, or nomadism, in which individuals move erratically between breeding and wintering areas without necessarily returning to their starting point the next year. Because this behavior is driven largely by food scarcity, the proportion of the population that wanders in any given winter varies from year to year. Juveniles are generally more nomadic than adults, and in some species one sex tends to wander more than the other.

Among the most nomadic North American owls are boreal, great gray, northern hawk, short-eared and snowy owls. All are specialist predators of small mammals that undergo regular population fluctuations, and the owls' wandering ways appear to be linked to these cycles. When prey populations are booming, the owls stay put. When they go bust, these specialist hunters cannot simply switch to other prey, so they are forced to look farther afield. Although this general principle is well understood, biologists are still trying to work out the exact relationship between prey abundance and the movement patterns of various owl species in different regions. What they do know is that large numbers of these owls periodically show up in areas south of their normal range in winter. These mass movements are known as irruptions or invasions.

Boreal owls are generally year-round residents with stable home ranges, but they will disperse, especially the females, when food is scarce. The boreal owl's main prey is the red-backed vole. Recent research by a team of biologists from Quebec shows that, at least in eastern North America, red-backed vole populations rise and fall on a four-year cycle, and boreal owl irruptions correspond to years when this rodent's populations are at their lowest.

Northern hawk owls are true nomads, breeding temporarily in areas with abundant prey and moving on when local food supplies decline. Widespread prey scarcity within the species' usual breeding range can provoke major southward movements. Historical records suggest that, in North America, both ten-year snowshoe hare population cycles and three- to five-year vole cycles (in this case, species such as meadow, long-tailed and tundra voles) influence the timing of this species' irruptions.

Snowy owls are similarly nomadic, breeding where and when they find abundant prey. Most members of this species move south in winter, but their movements may be either migratory or irruptive. Periodic invasions of areas beyond the regular wintering grounds correspond roughly to four-year lemming cycles in the Arctic. Snowy owls that appear to the east and west of the northern Great Plains during irruptions are mainly immature nonbreeders.

Depending on the year and where they live, great gray owls can be resident, migratory or nomadic, with vole cycles apparently driving the latter patterns. Irruptions south of their normal range mainly occur east of Saskatchewan.

CHAPTER EIGHT

Species Profiles

Profiles of North American Owls

The geographic scope of this book encompasses Canada, the continental United States and those parts of Mexico that lie north of the tropic of Cancer (23° 27' north latitude). This dividing line was chosen because it marks the boundary between two vast regions, each characterized by a distinctive geography, flora and fauna: the Nearctic realm to the north and the Neotropical realm to the south.

The profiles in this chapter summarize the significant details of the lives and habits of North America's 23 owl species, with information arranged in the nine sections described below. Each profile begins with a short essay that tells a story about the species or highlights a unique aspect of its biology.

Appearance

The appearance section describes adult owls. Sizes are given in terms of body length, measured from the tip of the bill to the end of the tail. The range of lengths includes both sexes, unless otherwise indicated. Females are larger than males in all species except the burrowing owl. Males and females of most owl species have identical plumage, but gender differences are described where applicable. Except for the pygmy-owls and members of the genus *Aegolius,* juvenile plumage is generally only subtly different from adult plumage. In all plumage descriptions, markings referred to as bars are horizontal; streaks are vertical. When trying to identify owls, remember that ear tufts are not always visible.

Species List

Family Tytonidae

Genus Tyto
Barn owl (*Tyto alba*)

Family Strigidae

Genus Aegolius
Boreal owl (*Aegolius funereus*)
Northern saw-whet owl (*Aegolius acadicus*)

Genus Asio
Long-eared owl (*Asio otus*)
Short-eared owl (*Asio flammeus*)
Stygian owl (*Asio stygius*)

Genus Athene
Burrowing owl (*Athene cunicularia*)

Genus Bubo
Great horned owl (*Bubo virginianus*)
Snowy owl (*Bubo scandiacus*)

Genus Ciccaba
Mottled owl (*Ciccaba virgata*)

Genus Glaucidium
Colima pygmy-owl (*Glaucidium palmarum*)
Ferruginous pygmy-owl (*Glaucidium brasilianum*)
Northern pygmy-owl (*Glaucidium gnoma*)

Genus Megascops
Eastern screech-owl (*Megascops asio*)
Vermiculated screech-owl (*Megascops guatemalae*)
Western screech-owl (*Megascops kennicottii*)
Whiskered screech-owl (*Megascops trichopsis*)

Genus Micrathene
Elf owl (*Micrathene whitneyi*)

Genus Otus
Flammulated owl (*Otus flammeolus*)

Genus Strix
Barred owl (*Strix varia*)
Great gray owl (*Strix nebulosa*)
Spotted owl (*Strix occidentalis*)

Genus Surnia
Northern hawk owl (*Surnia ulula*)

Voice

Owls are often heard rather than seen. The most common and distinctive adult vocalizations, including primary songs and other calls used in territorial advertising and defense, are reviewed.

Activity Timing and Roost Sites

This section provides information on the time of night or day when the owls are most active and most likely to be seen or heard. It also describes typical roosting sites and behavior.

Distribution

The range map included with each profile gives a visual overview of where the species is found in North and Central America. Notes on distribution provide additional details about North American year-round, breeding and winter ranges and summarize global distribution for those species that also reside elsewhere.

Habitat

General habitat requirements, essential habitat characteristics and seasonal changes in habitat use are described. Elevational ranges are given for most species.

Feeding

The feeding section lists typical as well as less common prey and describes the hunting techniques employed by the species.

Abandoned woodpecker holes are used for nesting and roosting by several species of North American owls, including the elf owl.

Breeding

The breeding section includes information on nest sites, clutch size and egg-laying dates, the duration of the incubation and nestling periods, and the interval between fledging and dispersal. Unless otherwise noted, all species are monogamous and raise only a single brood each year.

Migration and Other Movements

This section describes seasonal migration patterns, nomadic wandering and periodic irruptive movements beyond the species' usual range.

Conservation

Conservation status is discussed in terms of regional, national and continental population trends and factors influencing those trends.

Barn Owl *Tyto alba*

Unlike most North American owls, male and female barn owls do not have identical plumage. Females are generally darker than males and have more and larger black spots on the undersides of their wings and breast. Overlap in these traits makes it impractical to try to sex a barn owl based on plumage alone, but it's safe to assume that the darkest, most heavily marked individuals in any population are females and that the palest and least spotted are males.

While barn owls don't need visual cues to sort out the sexes, plumage spots may convey other important information. According to a team of European researchers led by Swiss biologist Alexandre Roulin, there is good evidence that female barn owls use their spots to advertise good genes. Specifically it appears that the greater the proportion of a female's plumage that is covered by black spots, the more resistant her young will be to nest parasites such as *Carnus hemapterus*, a tiny bloodsucking fly.

Roulin and his co-investigators found that the nests of heavily spotted female barn owls were less infested by these flies and that the flies in those nests had lower reproductive rates than in the nests of lightly spotted females. To test whether the mothers were somehow directly responsible for reducing parasite abundance and fecundity, the researchers switched nestlings between nests. This cross-fostering experiment confirmed that the offspring of intensely spotted females were more resistant to the bloodsucking flies than the offspring of less spotted females, regardless of whether they were raised by their own mother.

It's unlikely that barn owls are consciously aware of the genetic advantage their progeny will gain from having a heavily spotted mother, but males do preferentially select such females as mates. As to the mechanism underlying this arrangement, the leading hypothesis is that individuals that carry parasite-resisting genes have more energy available to invest in production of ornamental traits such as spots, and that evolution favors the instinctive recognition of such symbols of genetic fitness.

Appearance

The barn owl's heart-shaped face sets it apart from all other North American owls. Other distinguishing features of this medium-sized owl (length 12.5–16 inches / 32–40 cm) include absence of ear tufts; small dark brown to black

Year-round

eyes; long, rounded wings; short tail; and long, sparsely feathered legs. In flight the feet (which have light gray toes) extend beyond the end of the tail. The facial disk is white, often tinged brown, and the bill is ivory colored. The ruff is golden buff, with a black border on the lower edge. Dorsal plumage is gold to buff with extensive black and white markings. The white to buff underparts may be either immaculate or marked with black spots varying in number and size. Females are generally darker and more spotted than males, but the sexes overlap in these traits.

Voice

The advertising call, a prolonged gargling scream repeated every 1 to 20 seconds, is given mostly by males and usually while in flight near the nest. Females sometimes give a more broken, husky version of this call. Other vocalizations include distress and warning screams; fast twitters uttered by males when delivering food, by females when feeding young and by either sex as part of courtship; a mellow purring given by males to invite nest-site inspection or by females to solicit food; and a snoring call given mostly by females and nestlings—with great persistence when they're hungry.

Activity Timing and Roost Sites

Barn owls are usually strictly nocturnal. By day they typically roost out of sight in a nest cavity or similar location or, in summer, in dense tree foliage. Occasionally they use more exposed roosts such as barn rafters. They often change roost sites several times a day. Mates often roost together outside of the nesting season, and some males continue to roost with the female during incubation and, less commonly, after the eggs hatch.

Distribution

The barn owl, one of the world's most widely distributed land birds, is found on every continent except Antarctica. In Mexico barn owls are widespread but most common in lowland areas. Their range covers much of the United States, but distribution is patchy. In Canada they are found only in extreme southern British Columbia and in the Niagara Peninsula.

Habitat

Barn owls forage in open habitats such as grasslands, farm fields, deserts and marshes. They sometimes live in and around urban areas if they find suitable nest sites and foraging habitat.

Feeding

Barn owls primarily eat rodents such as voles, cotton rats and pocket mice. Other common small mammal prey are shrews, bats and rabbits. Their diet may also include some birds and occasional reptiles, amphibians, insects, scorpions and crayfish. They hunt mostly by cruising slowly across open areas 5 to 15 feet (1.5–4.5 m) above the ground, but sometimes they hunt from a perch.

Breeding

Barn owls are usually monogamous and mate for life, but polygyny occurs occasionally. Pairs sometimes produce a second brood and occasionally a third. These cavity-nesters use a wide range of sites, both natural (existing holes in trees, cliffs and rock outcrops; caves; self-excavated burrows in soft banks) and human-made (including barn lofts, church steeples, other buildings and haystacks). They readily use nest boxes. Nest sites are often reoccupied for years by a succession of pairs.

The clutch size is usually 5 to 7 but may increase to 10 to 12 when food is abundant. Laying most commonly begins in late winter or spring; in areas with mild winters it can begin almost anytime. The incubation period is about one month. After their first flights, at about two months old, juveniles continue to roost in the nest for several more weeks. They are fed by their parents for three to five weeks after fledging, becoming independent and dispersing in mid to late summer.

Migration and Other Movements

Barn owls are mostly year-round residents, but some individuals that breed in the northeastern United States may migrate. Juveniles commonly disperse in all directions from their natal territory, traveling up to 960 miles (1,600 km) through fall and early winter to find breeding sites.

Conservation

Populations are declining in many parts of North America, most significantly in the American Midwest. Possible causes include loss of nest sites and foraging habitat, vehicle collisions and the lethal effects of pesticides. In some areas, such as southern British Columbia, conversion of forested lands to farm fields allowed this species to expand its range, but these gains are now being reversed by urbanization.

Boreal Owl *Aegolius funereus*

In Europe *Aegolius funereus* is referred to as Tengmalm's owl, in honor of Swedish physician and naturalist Peter Gustaf Tengmalm. In North America it is known as the boreal owl, a name that reflects its residency across the continent's northern latitudes. It was not until 1963 that there was any scientific record of this species nesting south of Canada, but breeding is now well documented in the Rocky Mountains from Idaho to northern New Mexico and in subalpine areas of eastern Washington and Oregon.

The late discovery of breeding boreal owls in the United States led some ornithologists to speculate that they were recent arrivals, but close examination of the evidence suggests that they were there all along and had been overlooked. Is it really possible to miss an owl whose song can be heard up to 2 miles (3.5 km) away and that sometimes sings for two to three hours straight, taking only brief intermittent breaks? Apparently it is, if the owl lives in remote locales and does most of its singing at a time of year when few people are around.

Surveys conducted since the 1960s have established that, in southern portions of their North American range, boreal owls inhabit high-elevation forests of subalpine fir and Engelmann spruce or transition forests within 330 feet (100 m) of these subalpine habitats. In Idaho and Montana 75 percent of breeding occurs above 5,200 feet (1,580 m), while in Colorado most nest sites are above 10,000 feet (3,050 m). From February to April, when male boreal owls are most vocal, these areas remain snowbound and inhospitable to humans.

Although 19th- and early-20th-century naturalists and ornithologists were a hardy lot, they seldom ventured into prime boreal owl habitat in the western United States any earlier than June or later than October. Even their summer visits were short, and by then the nesting owls were relatively quiet. Nowadays paved roads, snowplows and high-tech skiing and cold-weather camping equipment make winter access to the boreal owl's high mountain haunts relatively easy, which is why the number of boreal owls seen and heard south of the border has increased so dramatically.

■ Year-round

Appearance

This species shows the greatest degree of size dimorphism of any North American owl, with females (length 10–11 inches / 25–28 cm) being noticeably larger than males (length 8–10 inches / 21–25 cm). Boreal owls have a

whitish facial disk framed by a black border, with white eyebrows, gray mottling toward the sides, relatively small yellow eyes and a buff-white to bluish yellow bill. Numerous small white spots are sprinkled across the dark olive brown forehead and crown. The back, wings and tail are also olive brown and are marked with larger white spots, including three rows on the tail. The creamy white underparts are broadly streaked brown to russet, and the densely feathered legs and feet are white with brown to russet mottling.

Voice

Male boreal owls have two songs. The primary song is a series of 11 to 23 notes delivered at a rapid, even rate; this approximately 2-second-long trill is sung repeatedly with only a few seconds between trills, often for up to 20 minutes at a time and sometimes much longer. The second song, a softer and longer trill, lasts up to a minute and is given when the female is near the nest. Males also utter a 4- to 10-note trill when delivering food to their mate or offspring. Female calls, usually given in response to male vocalizations, include a soft peeping and a harsh *chuuk*. A loud, hoarse screech, likely a contact call, may be uttered by either sex.

Activity Timing and Roost Sites

Boreal owls are mainly nocturnal except at high latitudes, where there is little darkness in summer. At the southern end of their range they occasionally hunt during daylight hours, more frequently in summer than in winter.

They roost on the branches of trees, often co-nifers, within 6 inches (15 cm) of the trunk. Individuals rarely use the same roost twice.

Distribution

This species inhabits northern latitudes across Eurasia and North America. Its North American breeding range encompasses the boreal forest zone from Alaska to Newfoundland, mostly north of the Canada–U.S. border, but dipping south into the extreme north of Minnesota and probably other eastern states. It also breeds in subalpine forests in the Rockies and other western mountains from northern British Columbia to northwestern New Mexico. Winter and breeding distribution differ only during periodic winter irruptions, when boreal owls may move a short distance south of their usual range.

Habitat

Boreal owls inhabit mostly mixed conifer-ous/deciduous boreal forest and western sub-alpine coniferous forests. In eastern North America they occasionally breed just south of the boreal, in deciduous and mixed forests. During nomadic movements they sometimes use other habitats. Mature or old-growth forest provides the best foraging habitat, but they often hunt in clear-cuts and farm fields just after the snow melts in spring.

Feeding

Boreal owls prey primarily on small mammals, especially voles and mice, but also shrews, pocket gophers, tree and flying squirrels, and chipmunks. They occasionally eat birds (captured on roosts or nests), insects (particu-larly crickets) and larger mammals up to the size of a juvenile snowshoe hare. They hunt from perches about 5 to 10 feet (1.5–3 m) above the ground.

Breeding

In Europe when vole populations peak, some male boreal owls (and a few females) take multiple mates; neither polygamy nor polyan-dry has been documented in North America. Boreal owls nest in cavities made by northern flickers or pileated woodpeckers or in natural tree cavities. They also readily use nest boxes. The usual clutch size is two to five. Laying be-gins between late March and late May and the incubation period is about one month. The young fledge when 27 to 36 days old. They are fed by their parents until they become independent, about three to six weeks after leaving the nest. Little is known about juvenile dispersal.

Migration and Other Movements

Boreal owls are nonmigratory, but individuals, especially females, often wander nomadically during years when food is scarce. Nomadic movements are linked to periodic winter ir-ruptions south of the breeding range.

Conservation

Although long-term trends for North American boreal owls are not known, population declines due to habitat loss are suspected. Logging of mature and old-growth forests eliminates nest trees and foraging habitat and often reduces the abundance of important prey species.

Northern Saw-whet Owl
Aegolius acadicus

The first scientific account of the northern saw-whet owl's migratory tendencies was published in *Auk* in 1912. Among evidence discussed by the article's authors, P.A. Taverner and B.H. Swales, was a report from ornithologists working at Long Point on Lake Erie, where "Saw-whets were at times captured in numbers by stretching old gill nets across the roads in the woods." Nearly a century later the work goes on, though specially designed mist nets have replaced the fishing gear and the owls are banded and

Year-round
Nonbreeding

released instead of being killed to become museum specimens. Today more northern saw-whet owls are banded annually than any other owl species in North America.

Nighttime mist-netting efforts during fall migration began in earnest in the 1960s. By 1997 more than 56,000 northern saw-whet owls had been caught and banded, nearly all of them in the Great Lakes region and the northeastern states. Although many questions remain unanswered, biologists have begun to gain a reasonable understanding of migration routes and behavior for saw-whets in the eastern half of the continent.

In the West, research on northern saw-whet owl migration is in its infancy; banding studies were started in 1999 in southwestern Idaho and north-central Oregon and in 2002 at Rocky Point, on the southern tip of Vancouver Island, British Columbia. So far, at all of these stations combined, only a few thousand saw-whets have been banded and examined to determine their age, sex and condition, but these efforts have significantly advanced scientific knowledge about the species. At Rocky Point 210 saw-whets were banded during the first year of operations alone, almost as many as had been banded in the entire province between 1956 and 2001.

In eastern North America this species is an irruptive migrant, meaning that there are large fluctuations in annual numbers of migrating individuals. On Virginia's Delmarva Peninsula, for example, banders netted 1,002 saw-whets in 1995 and 700 in 1999, but only 22 to 105 in each of the other years between 1994 and 2000, despite fairly consistent trapping efforts. Whether migration patterns for western saw-whets are comparable remains to be seen.

Appearance

The northern saw-whet owl is a small, short-tailed owl (length 7–8.8 inches/18–22 cm) with no ear tufts. Its dark brown dorsal plumage is marked with white streaks on the crown and white spots on the nape, back, wings and tail. The white underparts are broadly streaked reddish brown and the densely feathered legs and feet are white to buff. (The subspecies that inhabits Canada's Queen Charlotte Islands has buffier underparts and buff dorsal spots.) The facial disk is pale buff edged with dark brown around the outer edges, fading to white above, below and between the eyes, which are yellow to golden. The bill is black.

Voice

"Saw-whet" refers to the sound of saw teeth being filed, but there is debate about which of this owl's vocalizations inspired its name. Some say it was the male's advertising call—a series of short whistled notes normally repeated at a constant rate of about two per second, but sometimes more rapidly, such as when responding to a taped playback or when approaching the nest with food. Females sometimes give a softer, less consistent version of this song during courtship. Another candidate is the loud, sharp, squeaky *ksew* or *skriegh-aw*—a contact call given by both sexes during the breeding season, usually in sets of three, and during migration. Another contact call that suggests saw-sharpening is a nasal whine that resembles the advertising song in pitch but lasts only about two to three seconds.

Activity Timing and Roost Sites

Northern saw-whet owls are almost entirely nocturnal, but they occasionally hunt during daylight hours in winter. By day they roost in dense vegetation. The typical roost site is in a large tree (preferably a conifer), near the end of a low branch sheltered by overhanging foliage. Other common sites include next to the trunk of a small tree or in a dense shrubby thicket. Tree cavities are rarely, if ever, used for roosting. Nonbreeding birds typically roost in the same site for weeks or months, while breeding males usually use a different roost each day.

Distribution

The northern saw-whet owl breeds in south coastal Alaska, across much of southern Canada, through much of the northeastern and western United States and in parts of Mexico's interior south to Oaxaca. Uncertainty remains about the northern boundary of the breeding range. The limits of the winter range are also poorly defined, partly because they likely vary from year to year. Northern saw-whet owls are present year-round within the breeding range, except perhaps along the northern edges and at higher elevations in the Appalachians and western mountains. They also winter in southern parts of the Canadian Prairie provinces and in most parts of the United States that are outside the breeding range.

Habitat

Northern saw-whet owls breed in all types of forests and woodlands within their range but favor mature or old-growth stands. The highest breeding densities are found in coniferous forests. Forested riparian habitats are favored in many areas. Winter habitat varies widely and includes rural and semi-rural habitats, provided they offer suitable roosting sites and foraging habitat. In mountainous areas of the southern United States and Mexico this species is restricted to elevations between about 5,900 and 11,550 feet (1,800–3,500 m); elsewhere it is found down to sea level.

Feeding

Northern saw-whet owls primarily eat small mammals up to the size of juvenile pocket gophers, chipmunks and squirrels. Deer mice are the dominant prey across much of this owl's range, with voles contributing significantly to their diet in areas where they are abundant. These owls eat only small numbers of birds, mostly songbirds caught while migrating at night. They also eat some insects such as beetles and grasshoppers. Members of the Queen Charlotte Islands subspecies consume large quantities of intertidal invertebrates. Northern saw-whet owls hunt from low perches such as branches, shrubs or fence posts, along forest edges or in open habitat.

Breeding

Although usually monogamous, northern saw-whet owls are occasional polygynous when food is abundant. For nesting they depend on cavities made by woodpeckers (usually pileated woodpeckers or northern flickers), using natural tree cavities only very rarely. Nest boxes are readily accepted. The clutch size is usually five or six, sometimes four or seven. Egg laying begins between late February and April and the incubation period is 27 to 29 days. The young fledge when about 33 days old and are fed by the male or by both parents for at least one month, possibly up to two.

Migration and Other Movements

In eastern (and perhaps western) Canada and the United States, some northern saw-whet owls remain within the breeding range year-round, but many move south in fall. In the East the number of migrants varies widely from year to year; in irruptive years birds hatched that spring greatly outnumber adults among the migrants. Individuals that winter in the nonbreeding range are primarily females. It is unclear whether these patterns apply in the West. In Mexico this species is nonmigratory.

Conservation

Population trends are not known, but numbers are probably slowly decreasing in many areas. The greatest threat is logging of the mature and old-growth forests that best meet this species' breeding habitat requirements.

Long-eared Owl *Asio otus*

Rodents are an important part of the diet of many owls, including the long-eared, but not all rodents are equally vulnerable to these nocturnal hunters. Kangaroo rats and kangaroo mice in particular are well adapted to avoiding owl predation. The secret to their success is a superior ability to detect approaching owls and to dodge their attacks.

Kangaroo rats and mice both have exceptionally keen hearing in the low- to midfrequency ranges because their auditory bullae—the bony capsules in the skull that contain the middle and inner ears—are very large relative to their body size. They also have elongated hind legs and a kangaroo-like style of locomotion, which is more effective than a four-legged scamper when trying to make a fast getaway. The superb hearing and reflexes of kangaroo rats were confirmed by a researcher who used red light to watch barn owls and eastern screech-owls hunt these mammals in the dark. Just as an owl swooped down and was about to strike, the kangaroo rat would make a spectacular vertical leap, rocketing about 18 inches (46 cm) off the ground and landing about a foot (30 cm) away from its starting point. It would then hop away with no apparent sense of urgency. Surprisingly, the thwarted hunters made no further attempts to capture the fugitives, even though these owls ordinarily pursue prey that they miss on their first attempt.

A field study of long-eared owls in the Great Basin desert in Nevada offers further evidence that kangaroo rats and mice are elusive prey. Biologist Burt P. Kotler found that the owls' eating habits did not reflect the relative abundance of different prey species living in the sand dunes of the study area. Few kangaroo rats and kangaroo mice were killed by the owls, even though these rodents spent much of their time out in the open. Conversely, rodents that lacked these species' anti-predator adaptations—such as pocket mice, deer mice and harvest mice—rarely crossed the exposed ground between clumps of spiny greasewood, but were captured by the owls in high numbers.

Year-round
Southern limits of wintering range

Appearance

The long, close-set ear tufts for which this medium-large owl (length 14–16 inches / 35–40 cm) is named are prominent on perched

birds but barely discernible in flight. Long-eared owls have yellow to golden eyes surrounded by elongated black patches. White eyebrows and a white patch at the base of the black bill contrast with the buff to rusty facial disk. The dark brown ear tufts are edged with light rust and white. The whitish to buff underparts are boldly streaked and barred dark brown. The dorsal plumage is dark brown with whitish to buff mottling. Buff wing patches and black wrist markings on the underside of the wings are visible in flight. The densely feathered legs and feet are buff to rust. Females are distinctly darker overall than males, with richer rust coloring on the face and underparts, especially the thighs.

Voice

The male's advertising song is a series of 10 to more than 200 evenly spaced low *hoo* notes uttered every two to four seconds; delivery rate and pitch vary between birds. It is usually sung from a high tree perch and occasionally in flight or from the ground. Females call from the nest or nearby perch with a soft, nasal *shoo-oogh* repeated every two to eight seconds. Pair duets are based on these vocalizations. Alarm is most commonly expressed with a barking call. This species is mostly silent outside the breeding season.

Activity Timing and Roost Sites

Long-eared owls are usually strictly nocturnal but sometimes begin hunting before sunset, especially when feeding nestlings. By day breeding males roost in trees near their nests, usually singly but occasionally with other males whose nests are nearby. During the nonbreeding season long-eared owls roost in dense thickets or tree groves beside open foraging areas, either alone or communally with up to a hundred other owls (usually 20 or less), often perched less than 3 feet (1 m) apart. Winter roosts are sometimes shared with short-eared owls.

Distribution

The long-eared owl's breeding range covers a wide swath of North America, Eurasia and northwestern Africa between the Arctic Circle and the tropic of Cancer. Within North America these owls breed as far north as the southern Yukon and Northwest Territories and as far south as northern Baja California and Nuevo León. They winter from southern Canada to Oaxaca and Veracruz in Mexico.

Habitat

Long-eared owls nest and roost in dense or brushy vegetation such as stands of coniferous or deciduous trees or willow thickets. They hunt in open habitats such as grasslands, meadows, open forests and deserts. They inhabit areas from near sea level to more than 6,600 feet (2,000 m).

Feeding

Long-eared owls primarily eat small mammals, concentrating on those that are locally abundant. Voles dominate their diet in much of North America. Other favored prey include deer mice (in some Midwestern and Great Plains states), pocket mice and kangaroo rats (in arid parts of the West) and pocket gophers (in northeastern Oregon). They occasionally eat small birds, bats and reptiles. They mostly hunt on the wing, coursing back and forth close to the ground and seizing prey from the ground or from low vegetation. They sometimes hunt from perches, especially during windy weather.

Breeding

This species is usually monogamous. Males occasionally attempt polygyny, but one or both nests usually fail. One case of serial polyandry has been documented. Long-eared owls sometimes nest in loose colonies with nests as close as 46 feet (14 m) apart. They typically use stick nests made by other birds but sometimes nest on dwarf mistletoe clumps and natural platforms in saguaro cacti, and occasionally on the ground. Unusual nest sites include tree or cliff cavities and squirrel nests. The eggs (from 2 to 10, but 5 on average) are laid between mid-March and mid-May and incubated for 26 to 28 days. The young leave the nest when 21 days old but are flightless until about day 35. They are fed by both parents until 6 to 8 weeks old, then only by the father, who deserts them at 10 to 11 weeks. Soon after, the juveniles leave the nest area, traveling many miles within days of departing.

Migration and Other Movements

Long-eared owls that breed in northern Canada are believed to be regular migrants. Those breeding farther south are largely sedentary, but some wander nomadically. The long-distance record of 2,400 miles (4,000 km) is held by an owl that flew from Saskatchewan to Oaxaca. Others have traveled more than 1,800 miles (3,000 km), from Minnesota and Montana to central Mexico.

Conservation

Long-term population trends for long-eared owls are difficult to determine because of nomadism and local fluctuations in response to prey population cycles. There is evidence of declines in some parts of North America. One major threat is loss of riparian woodlands and isolated tree groves, especially in arid regions. Another is loss of foraging habitat, such as grasslands and marshes, to urban and rural housing developments. This species is listed as endangered, threatened or of special concern in a number of American states.

Short-eared Owl *Asio flammeus*

The short-eared owl, one of the world's most widely distributed owls, may soon add Florida to the long list of places where it nests. Historically this species was found in the southern United States and northern Mexico only during the nonbreeding season, but over the past three decades short-eared owls have started to show up in southern Florida in spring and summer. Rather than being stragglers that failed to migrate north after the winter, they seem to be prospective colonizers from across the ocean.

Until the late 1970s short-eared owls were rarely seen in Florida, and then almost invariably between early October and late March. Since then the number of sightings has increased and most have been seen in spring and summer. When at least five of these owls were spotted in southern Florida and another

was plucked from the Gulf of Mexico about 66 miles (110 km) west of Hernando County, all in the spring of 1994, ornithologists began wondering what was going on. After carefully examining a number of dead short-eared owls that had been collected in southern Florida between 1978 and 1998, as well as photographs of others, a trio of researchers concluded that most had come from the Greater Antilles, with Cuba being the most likely point of origin. Notable distinctions between Antillean short-eared owls and those belonging to the North American subspecies include differences in wing, tail and leg lengths, plumage coloring, foot feathering and body weight.

The invasion of the United States by Cuban short-eared owls was perhaps inevitable, since this species is renowned for traveling great distances over water. A short-eared owl once landed on a ship in the Pacific 653 miles (1,088 km) from land, and the presence of breeding populations on remote oceanic islands suggests that such epic journeys are not uncommon. If Antillean short-eared owls do eventually start nesting in southern Florida they will be following a well-established trend. Since 1932 Antillean land birds have been colonizing southern Florida—appearing first as vagrants, then increasing in number and ultimately breeding—at the rate of about one species per decade.

Appearance

The short-eared owl's small, centrally positioned ear tufts are generally invisible except in defensive postures. This species is 14.8 to

Breeding
Year-round
Nonbreeding

15 inches (37–38 cm) in length and has a large, round head. The large whitish and buff facial disk is set off by a white, brown and buff border and forehead and a white chin. Round black patches accentuate the yellow eyes, and the bill is black. The dorsal plumage is mottled dark brown and buff. The underparts are whitish to rusty with coarse dark brown streaks on the throat, chest and breast and thinner, sparser streaks elsewhere. The feathered legs and feet are white to buff. Buff wing patches are conspicuous in flight and black wrist patches are visible on the underwing. Males and females have similar plumage, but females generally have darker dorsal coloring and rustier, more heavily streaked underparts.

Voice

Males broadcast their primary song—a rapid series of 13 to 18 short toots—during courtship flights or, less often, from the ground or an elevated perch. Females sometimes respond with a *keeeyup* call. Both sexes utter various alarm calls that resemble barks, screams and whines. The raspy, high-pitched bark is this owl's most frequently heard call and usually the only vocalization heard outside of the breeding season.

Activity Timing and Roost Sites

These owls hunt during either day or night and tend to be most active around dusk and dawn. The amount of daytime hunting depends on the season, the weather, prey availability and the food demands of dependent offspring. Males perform courtship flights at all hours. Short-eared owls usually roost on the ground, sometimes on slightly elevated sites such as abandoned vehicles or stump piles. During the nonbreeding season they often roost communally in areas of dense grass. Occasionally they roost in trees, either alone or with other short-eared owls or long-eared owls.

Distribution

The short-eared owl is found throughout much of North America and Eurasia, as well as in parts of South America and Africa and on numerous major islands. Its North American breeding range encompasses Alaska and nearly all of Canada, except some Arctic islands, and extends across the northern United States from central California to Virginia. Its North American winter range includes the southern edges of several Canadian provinces, most of the United States and the northern half of Mexico (mainly the northwestern area, including Baja California). Distribution is patchy within both breeding and winter ranges.

Habitat

During the breeding season short-eared owls occupy large expanses of open country, including prairie and coastal grasslands, tundra and subalpine meadows. In winter they use similar habitats, as well as large open areas such as marshes, weedy or stubble-covered fields, extensive forest clearings, gravel pits and shrub thickets.

Feeding

In North America short-eared owls eat mainly small mammals, particularly voles. Other common mammal prey are mice, rats, lemmings, shrews, moles, pocket gophers and rabbits. Birds, including shorebirds, songbirds, terns and gulls, make up a small proportion of their diet; avian prey are more important in coastal areas than inland. When hunting they generally fly close to the ground but sometimes hang in the wind or hover at greater heights. Occasionally they use a pole or a hill as a hunting perch.

Breeding

Short-eared owls are loosely colonial ground-nesters. Their nest is a shallow bowl scraped out by the female and sparsely lined with grass and downy feathers. Typical nest sites are dry (often on a small knoll, ridge or hummock) and surrounded by enough vegetation to conceal the incubating or brooding female. The clutch size increases with latitude and varies with food availability; the average number of eggs is 6 and the range is 1 to 11. The eggs are laid from late March to June, depending on latitude, and incubated for 26 to 31 days. The young leave the nest on foot 14 to 17 days after hatching and can usually fly when about a month old. The duration of parental care is not known.

Migration and Other Movements

Except in a few areas, the Canada–U.S. border roughly marks the dividing line between migratory and nonmigratory populations. It is unclear whether adults are regularly nomadic, but dispersing juveniles are known to travel long distances. Cross-country movements of up to 1,135 miles (1,891 km) have been recorded. During prey population irruptions, large numbers of short-eared owls may concentrate in areas where food is plentiful.

Conservation

This owl's status is difficult to assess because of its nomadic tendencies and short-term local population fluctuations in response to varying prey abundance. Breeding populations in the Canadian Prairie provinces and some western states appear to have declined. Populations are stable or increasing in the Dakotas, northeastern Minnesota and eastern Montana and Wyoming. In much of the northeastern United States breeding has been significantly restricted. Declines are due largely to the encroachment of agriculture and urban and recreational developments on nesting and foraging areas, leading to habitat loss or disturbance by humans, livestock and pets.

Stygian Owl *Asio stygius*

In 1832 German zoologist Johann Georg Wagler penned the first scientific description of a *styxeule*—a bird we know today as *Asio stygius,* or the stygian owl. These names all refer to the river Styx of Greek mythology, which separates the realm of the living from the underworld inhabited by the souls of the dead. In Brazil, the country Wagler's original specimen came from, this is the Devil's owl—*coruja-diabo*—so named because of its long ear tufts, almost black plumage and eyes that glow bright red when illuminated at night.

Until the 1990s stygian owls were not thought to occur north of Mexico. In December 1996 three members of the Wright family from North Carolina noticed a gray hawk mobbing an unusual owl in Bentsen–Rio Grande Valley State Park, near Mission, Texas. Despite harassment by the hawk, the owl remained perched in the same tree all day, offering excellent views to the handful of lucky birders and photographers who were in the area. By the time it flew off at dusk it had been declared the first stygian owl on record in the United States. As it turned out, however, there was a previous record. Two years earlier in the same park, experienced birder Mel Cooksey had photographed what he judged to be an odd-looking long-eared owl. After the Wright sighting he reexamined his old photos and realized the bird's true identity.

The stygian owl's known distribution in Mexico comes nowhere near the area where these owls were seen, but this species' vast and discontinuous range, which includes Cuba and Hispaniola, suggests that it is an exceptionally good disperser.

Appearance

This medium-large owl (length 15–18 inches / 38–46 cm) is distinguished by its dark plumage, long, close-set ear tufts and dark yellow eyes. A whitish, somewhat diamond-shaped patch on the forehead and another whitish patch at the base of the black bill contrast with the charcoal black facial disk and dark ear tufts. The dorsal plumage is blackish brown, sparsely spotted and barred in buff. The underparts are dusky buff, heavily streaked and barred with dark brown. The legs are feathered and the toes bristled.

Voice

The male's territorial call is a deep, loud *woof* with a descending inflection. In northern Mexico it is repeated every 6 to 10 seconds; in some other parts of this owl's range the interval is shorter. The female's short, shrill, catlike call is associated with pair interactions. A long series of scratchy notes is given by both sexes when excited.

Year-round

* Rare visitor

Activity Timing and Roost Sites

Stygian owls are largely or entirely nocturnal. During the day they usually roost in dense vegetation.

Distribution

The stygian owl's extensive breeding range runs discontinuously from northern Mexico to northern Argentina and includes some Caribbean islands. In Mexico the known range follows the Sierra Madre Occidental from southern Chihuahua to Jalisco; it is discontinuous along the Pacific and Atlantic slopes in the south. This species is a rare winter visitor to southeastern Texas.

Habitat

In Mexico stygian owls are found at elevations of 4,950 to 9,900 feet (1,500–3,000 m) in pine and pine/oak forests or dense cloud forest. Their sporadic appearances in the United States have been in the Rio Grande floodplain forest.

Feeding

Small birds (taken while roosting at night) and bats (caught in flight) appear to constitute a significant part of the diet. Stygian owls also eat small mammals and insects. They generally hunt from perches.

Breeding

Very little is known about the stygian owl's breeding biology, especially in Mexico. It seems to nest mostly in trees, using abandoned stick nests made by other birds; one nest of shredded palm leaves on the ground was found in Cuba. The reported clutch size is two.

Migration and Other Movements

Stygian owls are nonmigratory. Dispersal has not been studied, but the size and patchiness of this species' range suggest that at least some individuals, probably juveniles or nonbreeding birds, at times make long-distance movements.

Conservation

Stygian owls are common to fairly common in northwestern Mexico. They have become very rare in the Caribbean because of deforestation and killings by people who associate them with witchcraft.

Burrowing Owl *Athene cunicularia*

Before Europeans arrived in North America, the Great Plains biome covered about 15 percent of the continent, and prairie dog colonies were spread across more than 387 million acres (155 million ha) of these vast grasslands. Today the prairie dog's native habitat has been reduced to less than 1 percent of its original extent, causing precipitous declines in populations of these mammals, which were also shot and poisoned by the millions throughout the 20th century. Although burrowing owls use holes dug by a variety of animals, and in Florida often dig their own, the prairie dog was and still is their key housing agent in most parts of North America. For these owls the double blow of losing so much habitat and the majority of their primary burrow providers has been devastating.

Because most burrowing owl habitat in North America is privately owned, voluntary stewardship by landowners is critical to this species' survival. One of the leaders on this front is Saskatchewan's Operation Burrowing Owl, which was begun in 1987. This program has enlisted the support of more than five hundred ranchers, farmers and other landowners and has led to conservation measures being implemented on more than 50,600 acres (20,235 ha) of private land. Although much of this land is not currently occupied by burrowing owls, these measures preserve it as potential habitat that may be recolonized.

Operation Burrowing Owl encourages landowners to take actions such as continuing to graze cattle, returning marginal croplands back to pasture, tolerating burrowing mammals in pastures, avoiding or limiting pesticide use, maintaining some areas of unmowed grass to increase burrowing owl prey, and working around active nest sites only after the owls have moved south. People who live near nesting burrowing owls—whether they are landowners or their neighbors—are advised to keep domestic cats indoors and to "give the owls a brake" by driving more slowly when passing known breeding areas. Members of the public elsewhere can do their part by supporting pesticide-free agriculture and buying free-range beef.

Breeding

Year-round

Nonbreeding

These young burrowing owls have not yet gained their adult plumage.

Appearance

Long legs, sparsely feathered on the lower portion, are a distinguishing feature of this relatively small, short-tailed owl (length 7.5–10 inches/19–25 cm). Burrowing owls lack ear tufts and their heads appear wide and oval. Bright yellow eyes and a cream to greenish yellow bill are set in a brownish facial disk framed by prominent white markings above and below. A dark brown half-collar separates the white chin and throat from a white chest band. The breast, abdomen and flanks are broadly barred in brown and buffy white. The under-tail area and upper legs are white to beige, and the feet and unfeathered part of the legs are dark gray. The dorsal plumage is a rich sandy brown with buffy white markings: streaks on the head, and spots and bars on the back, wings and tail.

Voice

Burrowing owls vocalize mainly during the breeding season and near the nest burrow.

The male's primary song, used in courtship and territorial defense, is a two-note *coo-oooo* given while bowed, with the body nearly parallel to the ground and white facial markings exposed. This song, with or without an additional warble at the end, is also given during copulation. Females have two copulation calls, the most common being a series of down-slurred notes. Females have several vocalizations associated with nest defense and food solicitation, including the *eep* call and the rasp. Both sexes utter single *chucks*, multi-note chatter calls and screams as warnings and when mobbing predators.

Activity Timing and Roost Sites

Burrowing owls hunt most intensively at dusk and dawn, but also at other times of day and night. Daytime activity occurs mainly during the breeding season. During incubation and brood rearing the male perches near his burrow throughout the day. Once nestlings are three to four weeks old, they sometimes stand and sun themselves near the burrow entrance. They also stand at the entrance when being fed by their parents, often running into the burrow to eat, then emerging a few seconds later. Burrowing owls usually roost in the mouth of the nest burrow or a satellite burrow or in a depression in the ground. In Florida they occasionally roost in shrubs or trees.

Distribution

Burrowing owls are found from southern Canada to the tip of South America, as well as in parts of the West Indies. There are two subspecies in North America. The Florida subspecies is a year-round resident there (and in the Bahamas). The breeding range of the western subspecies includes southern parts of the Prairie provinces and British Columbia and much of the western United States and northern Mexico. Individuals are occasionally seen as far north and east as Ontario and New York. The winter range of the western subspecies overlaps the breeding range in the southern United States and northern Mexico and extends south into nonbreeding areas of Mexico and Central America.

Habitat

Burrowing owls breed in dry, open areas with short vegetation or bare ground, including human-modified habitats such as golf courses, airports and cemeteries. The presence of existing nest burrows is a critical requirement for the western subspecies; Florida burrowing owls usually excavate their own burrows. Little is known about winter and migration habitats of migratory individuals, but they probably resemble breeding habitat.

Feeding

As opportunistic hunters, burrowing owls have a seasonally variable diet generally dominated by invertebrates—such as scorpions, beetles, locusts, crickets and earwigs—in summer and small mammals and birds in winter. They also eat amphibians and reptiles. Hunting techniques include walking, hopping or running along the ground, flying from a perch, hovering over tall vegetation and catching aerial prey on the wing.

Breeding

Occasional polygyny occurs in Saskatchewan, but elsewhere only monogamy has been observed. Most burrowing owls in Florida and southern California remain paired year-round and mate for life. Migratory members of the western subspecies are solitary in winter and often change partners from year to year. Florida burrowing owls occasionally raise two consecutive broods in a year, and one instance of a second brood is known from southern California. Burrowing owls generally nest in loose colonies, with burrows as close as 15 yards (14 m) apart. They mostly use burrows excavated by other animals but sometimes excavate their own. Artificial sites include nest boxes, pipes and culverts and, very rarely, buildings.

Egg-laying dates for the western subspecies range from late March in the southern United States to mid-May in Canada. In Florida the main laying period is February to May. The maximum clutch size is 12, but usually 2 to 6 in Florida and 7 to 9 elsewhere. The incubation period is 28 to 30 days. Nestlings first venture out of their burrow at about 14 days old. During the next two weeks families may make one or more moves to satellite burrows. The young begin flying at four weeks and can fly well by six weeks. Parental feeding then tapers off and juveniles migrate or disperse when about three months old.

Migration and Other Movements

Most burrowing owls from Canadian and northern U.S. populations are migratory, leaving their breeding grounds in September and October and returning between March and May. They travel up to 180 miles (300 km) a night. Routes and wintering locations for specific populations are not well-known. Populations in Florida, southern California and northern Mexico are predominantly nonmigratory. In other areas some individuals may remain year-round.

Conservation

The burrowing owl is listed as endangered in Canada (where it could soon be extirpated if declines are not reversed) and threatened in Mexico. In the United States the status of the western subspecies varies regionally. Losses have been greatest along the eastern and northern edges of the subspecies' breeding range and in parts of Oregon and California. Some populations of the Florida subspecies have declined sharply because of loss of natural habitat. This subspecies has expanded its range in some areas where human activities have created new habitat, but these populations often collapse as development intensifies.

Threats include habitat loss, degradation and fragmentation; eradication of prairie dogs and other burrow providers; exposure to pesticides; killing, harassment or burrow damage by humans or domestic pets; and vehicle collisions. Attempts to reestablish extirpated populations through captive breeding and reintroduction have had minimal success.

Great Horned Owl *Bubo virginianus*

Great horned owls consume a greater diversity of prey than any other North American owl. Their diet includes every type of vertebrate—mammals, birds, reptiles, amphibians and fish—and a wide range of invertebrates, including earthworms, crayfish, scorpions, spiders, centipedes and both larval and adult insects. Quarry range in size from grasshoppers to great blue herons. However, although there is much geographical, seasonal and year-to-year variation in prey, mammals dominate the menu, constituting about 90 percent of this owl's diet in most regions.

Birds generally make up about a tenth of the diet, but great horned owls occasionally identify a source of avian prey that is easily exploited and focus their hunting attention there. Great horned owls living on Washington State's Protection Island dine almost exclusively on rhinoceros auklets in summer. An estimated 17,000 pairs of these burrow-nesting seabirds nest annually on the 368-acre (147 ha) national wildlife refuge, which is also home to a number of other bird species but few mammals, except for some shrews and the odd chipmunk. Of 129 great horned owl pellets collected by researchers on the island over two summers, 112 contained only auklet remains and another eight contained a mix of auklet and other bird bones. Mammal remains were completely absent.

The kind of opportunism exhibited by the Protection Island owls is likely responsible for this species' undeserved reputation as a ruthless poultry killer. An individual owl that discovers a poorly secured chicken coop might well return nightly for an easy meal, especially if there is a local scarcity of mammal prey, but the majority of great horned owls executed for this alleged crime have been innocent victims. Fortunately such killings are now rare, as well as illegal.

Appearance

Great horned owls are distinguished by their large size (length 18–25 inches / 46–63 cm), bulky shape, prominent ear tufts, large yellow eyes and white bib. Females are noticeably larger than males. Plumage coloring ranges from grayish to gray-brown to buffy brown. The darkest individuals are found in humid coastal areas and the lightest in the arid subarctic Prairies, Great Plains and American

■ Year-round

Southwest. The head and back are mottled whitish, dusky and dark brown. The paler underparts are mostly barred dark brown, except for the black-blotched upper chest and the bib—an immaculate white crescent across the throat. A narrow black border encircles the outer edges of the facial disk. The legs and feet are feathered.

Voice

The primary song is a series of deep, mellow, far-carrying hoots, with considerable variability in length and number; generally there are three to eight hoots in each set, but sometimes a single hoot is repeated at irregular intervals. Male hoots are more prolonged and elaborate, as well as lower-pitched. During courtship mated pairs often synchronize their territorial singing. This owl's diverse array of calls includes piercing screams (most commonly uttered by females, especially when defending the nest), sharp whistles (for contact with fledglings) and short, barking alarm notes, as well as chuckles, growls, hisses, squawks, catlike meows, soft coos and quavering cries.

Activity Timing and Roost Sites

Although mainly nocturnal, great horned owls sometimes hunt during daylight hours when food requirements are hard to meet, such as when they are feeding young. Their varied roosts include tree foliage (coniferous or deciduous), thick brush, tree cavities, cliff ledges and buildings. Males typically use a single preferred roost site during nesting and change roosts from day to day during the non-breeding season.

Distribution

The great horned owl has a more extensive North American breeding range than any other owl, encompassing nearly all of Canada and the United States, from the arctic tundra south, and much of Mexico, except parts of the humid southeast. It tends to avoid the Appalachian and Rocky mountains and the Mojave and Sonoran deserts. Great horned owls also inhabit parts of Central and South America.

Habitat

Habitats occupied by this species are extremely diverse. The minimum requirements are suitable nesting and roosting trees and sufficiently large hunting areas that are relatively open but with at least a few scattered perches, such as open woodlands, wetlands, grasslands, clearings, farm fields or pastures. The trees may be deciduous, coniferous or a mix of both, and elevations range from sea level to about 13,200 feet (4,000 m). Great horned owls occasionally take up residence in cities, especially around parks and golf courses.

Feeding

These opportunistic hunters kill a wide range of vertebrate and invertebrate prey but focus mainly on rabbits, hares, rodents and waterfowl. Great horned owls hunt primarily from perches (including trees, tall shrubs, rock outcrops, cliff ledges, bridges, poles and buildings) in open areas. They sometimes move on foot when hunting small mammals or invertebrates in grassy areas or other suitable habitat. They may also forage by flying low across open areas such as grasslands or sagebrush habitats.

Breeding

Although great horned owls are solitary outside the breeding season, members of a pair often share the same territory year-round and may mate for life. They nest mostly in stick nests made by other birds. They also regularly use cliff ledges and rock outcrops, especially in arid areas, as well as tree cavities and hollow broken-topped trunks. Unusual sites include abandoned buildings, bridge beams and the ground. They readily use nest baskets or platforms. The usual clutch size is two, increasing to a maximum of five when food is abundant. Laying dates vary widely with latitude—in Florida, rare very early clutches have been found in late November; in northern areas, laying is sometimes not completed until mid-May. The eggs are incubated for 30 to 37 days and the young leave the nest about six weeks after hatching. Fledglings remain with their parents through summer and gradually disperse before the next breeding season.

Migration and Other Movements

Although great horned owls are nonmigratory, members of northern populations periodically move long distances south in winter, particularly when snowshoe hare populations crash. The longest known journey of this type is 1,235 miles (2,058 km), from Alberta to western Illinois.

Conservation

Great horned owl populations are thought to be healthy in most areas, but long-term trends are not well-known. Densities are naturally low, with numbers fluctuating in relation to prey populations where these are cyclic. While illegal shooting, leghold traps, vehicle collisions and electrocution kill many individuals, their impact on populations may be minimal. These owls adapt well to habitat change, provided nest sites are available. They may benefit from clear-cut logging in densely forested areas.

Snowy Owl *Bubo scandiacus*

Although lesser and greater snow geese have a virtually unlimited selection of suitable nesting areas in the Arctic, they often choose to raise their families close to snowy owl nests. Since the owls return to the breeding grounds first and have already begun laying by the time the snow geese arrive, this neighborliness is clearly initiated by the geese. Researchers have found as many as 23 goose nests surrounding a single snowy owl nest, including one that was only 13 steps away from the owl's quarters.

Why would a potential prey species nest on the doorstep of a known predator? Because, as the saying goes, my enemy's enemy is my friend. Snowy owls seldom kill geese or goslings, preferring lemmings above all other prey, but Arctic foxes steal goose eggs at every possible opportunity and are the main cause of nesting failure for snow geese in many areas. Since the owls will attack and drive away any predator they consider a threat to their own eggs or young, geese that settle next to snowy owls gain an automatic security system.

Studies have shown that snow geese that nest within 600 yards (550 m) of a snowy owl nest have a much higher breeding success rate than those that nest farther away. And the greater the proximity, the better the protection. The geese seem to know this instinctively, as those that arrive on the breeding grounds earliest tend to claim the closest positions. When they can, some other types of waterfowl, such as eiders and brants, take advantage of snowy owl defensive behavior in the same way.

Appearance

These very large, predominantly white owls have relatively small golden eyes, a black bill (nearly concealed by facial feathers) and thickly feathered legs and feet. Mature males are either entirely white or only sparsely marked with narrow pale gray or brown barring on the breast, back, wings, head and tail, or on only some of these parts. Mature females have moderate to extensive barring on the breast, wings, head and tail. Immature owls are more heavily barred than adults—immature females have the most extensive markings and immature males resemble adult females. Snowy owls probably gain adult plumage at two years old, during summer. Although plumage characteristics are sometimes helpful for sexing snowy owls, there is no discernible difference between the most heavily barred adult males and the whitest adult females. However, females (length 24 to 28 inches/60–70 cm) are noticeably bigger than males (length 22–25 inches/55–64 cm).

Voice

Snowy owls, especially breeding males, are highly vocal around the clock during the breeding season. Wintering and nonbreeding birds are mostly silent. Territorial males deliver booming hoots from the ground or a perch or in flight. These can be heard more than 7 miles (11 km) away under favorable conditions and are typically answered by hoots from other males, both distant and nearby. The hoots are usually given in twos but sometimes in a series of six or more, with a one- or two-second pause after each one. Females rarely hoot. Both sexes, but mostly males, give a rattling bark when disturbed near the nest. The most common female vocalizations are a mewing call given before and after feeding by the male and during distraction displays, and a rapid, high-pitched *ke-ke-ke-ke-ke* given before copulation and when feeding nestlings.

Activity Timing and Roost Sites

During the continuous light of arctic summer, snowy owls hunt at all hours. When lemmings are plentiful, males may spend as much as

80 percent of their time resting or sleeping on hummocks, rock outcrops or other perches. Favorite roosts get well fertilized by owl droppings; they are visible as green patches of lush vegetation in an otherwise barren landscape. In winter, snowy owls often hunt during the daytime; whether they also hunt at night is not known.

Distribution

The snowy owl has a circumpolar distribution, breeding at high latitudes along the northern coastlines of North America, Greenland and Eurasia, and moving south in winter. In North

■ Breeding
■ Wintering

Breeding and wintering distribution of snowy owls in North America. Winter distribution from Christmas Bird Count data, 1952–81:
1 = regions reporting snowy owls in all winters
2 = > 50% of the winters
3 = < 30% of the winters
4 = only 2 winters

America snowy owls winter from the breeding range to southern Canada and the northern United States, with occasional forays as far south as central California and the Gulf States from Texas to Florida.

Habitat

Snowy owls nest on open tundra from near sea level to 990 feet (300 m) on inland mountain slopes. They favor high, rolling terrain with many hillocks for perches and nest sites. Foraging areas on the breeding grounds include wet coastal or inland meadows. In winter they frequent open, tundra-like habitats such as grasslands, farm fields, marshes, beaches and sand dunes.

Feeding

When lemmings are abundant, snowy owls eat little or nothing else. When they are scarce or absent, these opportunistic hunters consume a wide variety of prey, most commonly mammals ranging in size from small rodents to large snowshoe hares. Avian prey include waterfowl (from small grebes and ducks to medium-sized geese), ptarmigan and songbirds. Occasionally they eat fish, amphibians, crustaceans and carrion. Snowy owls use a variety of hunting techniques, including flying low and suddenly dropping down on prey, hovering about 50 feet (15 m) above the ground before striking, and pouncing on prey from a perch or while walking or standing on the ground. Ridges, hummocks or boulders serve as perches on the tundra. In winter habitats they also use fence posts, telephone poles and buildings. Prey are captured in the air, on the ground or on water.

Breeding

Snowy owls are usually monogamous, but both polygamy and polyandry occasionally occur. When food is scarce they do not breed. The female creates a shallow nest, located on a snow-free, flood-resistant site with a clear view, such as a windswept rise, a hummock, a frost-heaved patch of ground or a boulder. The clutch size varies with prey abundance: three to five when food is limited; seven to eleven when it is plentiful. Laying typically begins in mid-May and the incubation period is about 32 days. The chicks leave the nest two to three weeks after hatching but cannot fly well until at least seven weeks old; they are fed throughout this time and probably for another week or two.

Migration and Other Movements

Where and when conditions allow, some individuals remain within the breeding range year-round. However, most are migratory, moving in a generally southward direction in autumn. Winter numbers are generally highest on the northern Great Plains. During periodic winter irruptions, large numbers of snowy owls, mainly immature birds, move to areas well beyond the usual winter range. These irruptions are apparently linked to the lemming cycle, but other factors, such as weather, may also be influential.

Conservation

Snowy owl numbers fluctuate dramatically. Reproductive highs and lows are closely tied to lemming population peaks and crashes. Long-term population trends in North America are not known. Recent research showing some movement of snowy owls between Alaska and Russia suggests that conservation efforts should be coordinated internationally.

Mottled Owl *Ciccaba virgata*

In February 1983 Dan Hillsman spotted a dead owl on the road just outside the entrance to Bentsen–Rio Grande Valley State Park in Hidalgo County, Texas, and stopped for a closer look. It was not a species he recognized. Although he had no desire to linger over the carcass, which was none too fresh and had been attacked by ants, he paused long enough to take one photograph, with a coin placed beside the bird for scale.

Three years later Hillsman mentioned his puzzling find to fellow birdwatchers Greg Lasley and Chuck Sexton. After examining the color slide, Lasley and Sexton agreed with Hillsman's tentative identification. The dead bird appeared to be a mottled owl, even though this species was unknown north of Mexico. From there the photo made the rounds to five different experts, who studied it independently and agreed unanimously that the bird was most certainly a mottled owl, the first ever reported in the United States. Another two decades would pass before the first confirmed sighting of a live mottled owl in the United States. This intrepid individual spent several days in Hidalgo County in July 2006 and was seen on a day roost a number of times.

Ciccaba virgata, known as *el búho café* in Mexico, is probably the most abundant and widespread forest owl of the New World tropics and subtropics. At the northeastern edge of its range, in Tamaulipas and Nuevo León, it inhabits subtropical woodlands along streams that cut through semiarid thorn brushlands. These habitats have been extensively cleared throughout the Gulf coastal plain to make way for agriculture, but scattered fragments endure here and there.

One place that still has intact native habitat is the Sierra Picachos. This outlying range of the Sierra Madre Oriental in northern Nuevo León is visible on the horizon from Texas and is connected to Hidalgo County by a woodland corridor. Since mottled owls are known residents of the Sierra Picachos, these mountains may be the source of the cross-border visitors.

Appearance

This medium-sized owl has dark brown eyes and no ear tufts. Lengths range from 11.5 to 15 inches (29–38 cm); the smallest individuals live in northern Mexico and the largest in the Amazon region. Females weigh significantly more than males.

There are two color morphs. Light-morph mottled owls are found in northwestern Mexico and other relatively dry areas. They have a brownish facial disk with a narrow white border and bold white eyebrows and whiskers. The dorsal plumage is dark brown, flecked and finely barred whitish or pale buff, with a row

Year-round
* Rare visitor

of whitish spots in the shoulder area. The underparts are whitish to pale buff, mottled dark brown on the upper chest and streaked dark brown below. Dark-morph mottled owls mostly inhabit more humid and equatorial regions. Their overall coloring is darker and the whitish to pale brown eyebrows and whiskers are less striking. In both morphs the legs are feathered and the bare toes are gray to yellowish, as is the bill.

Voice

Mottled owls are highly vocal. The territorial call consists of 3 to ten 10 hoots uttered at approximately half-second intervals, often accelerating and becoming louder, then fading with the final one or two notes. Also in their repertoire are a longer, slower series of up to 9 hoots delivered over a 20-second period with gradually increasing volume and a string of up to 20 rapid hoots with increasing tempo but becoming softer and lower at the end. Single or double hoots are occasionally given, possibly as alarm calls. Other vocalizations include deep barks, which are surprisingly loud for this owl's body size, and a catlike yowl given by females, likely when soliciting food.

Activity Timing and Roost Sites

Mottled owls are strictly nocturnal. In Guatemala, the only place where their roosting habits have been studied, their daytime roosts are in densely wooded areas, where they perch on tree branches or among vines. They often spend the day within 6.5 feet (2 m) of the forest floor, particularly in very hot weather.

Distribution

The northern limits of the mottled owl's breeding range are reached in southern Sonora and northern Nuevo León. The two arms of the range follow the Pacific and Atlantic slopes south to meet in southern Mexico. Mottled owls are also found in much of Central America and northern South America. This species is a very rare visitor to Texas.

Habitat

Mottled owls occupy a wide variety of dense to semi-open treed habitats, including humid evergreen forest, dry lowland forest, thorn forest and pine/oak woodland. They are also found in human-modified habitats such as second-growth forest and coffee and cacao plantations, as well as around settled areas. In Mexico they are resident from sea level to 8,250 feet (2,500 m).

Feeding

The mottled owl's principal prey are large insects such as grasshoppers, cockroaches and beetles, but small rodents are an important part of its diet. These owls also occasionally eat other small vertebrates, including lizards, snakes, frogs, bats and birds. They typically hunt from low perches.

Breeding

There has been little study of this species, breeding habits. Mottled owls are apparently monogamous and usually nest in a natural tree cavity or the decayed top of a broken tree

trunk. Few clutches have been counted, but most contained two eggs. In Mexico, nests with eggs have been found from early April to mid-June, but the starting date for laying is not known. Incubation takes at least 28 days. The young are incapable of sustained flight when they leave the nest at about one month old. One study in Guatemala found that both parents roosted with and fed their offspring for at least three months after fledging.

Migration and Other Movements

Mottled owls are nonmigratory. As adults they appear to be sedentary, remaining on their home range year-round. Nothing is known about when or how far juveniles disperse.

Conservation

The mottled owl is fairly common to common in most parts of its range and seems to be fairly tolerant of human presence and limited habitat alteration.

Colima Pygmy-Owl
Glaucidium palmarum

For taxonomists—scientists who classify and name organisms—the pygmy-owls are one of the world's most confusing groups of owls. Twenty years ago taxonomists recognized only 15 different species within this genus. These days the number of officially named species has more than doubled, largely because of closer study of pygmy-owls living in New World tropical and subtropical regions. Among these "new" species is the Colima pygmy-owl, which gained a place on the American Ornithologists' Union's checklist of North American birds in 1997.

Before being elevated to full species status, the Colima pygmy-owl was considered one of eight subspecies of the least pygmy-owl. These eight are now divided among four species, one of which retains the original name. Besides the Colima pygmy-owl, the other newly defined species are the Tamaulipas pygmy-owl, found in northeastern Mexico to just barely north of the tropic of Cancer; the Central American pygmy-owl of southeastern Mexico, Central America and the Pacific slope of northern South America; and the least pygmy-owl of southeastern Brazil and adjacent parts of Paraguay.

The eight subspecies formerly known as *Glaucidium minutissimum* were originally lumped together because of similarities in appearance. While the reclassification takes plumage and body measurements into account, it is also based on information about vocalizations and ecology that was previously unavailable or not considered relevant. To determine the proper relationships within this complex group of owls, ornithologists Steve N.G. Howell and Mark B. Robbins listened to and analyzed tape recordings of 555 songs sung by 37 different least pygmy-owls. They also examined every one of the dead least pygmy-owls tucked away in museum drawers throughout North America—151 specimens in all. The oldest Colima pygmy-owl specimen was an adult female collected by E.W. Nelson and E.A Goldman on April 5, 1897. Their report on this previously undocumented species, which they called the Tepic pigmy owl, notes that she "was shot in the midst of a palm forest on a low ridge near the sea coast south of San Blas, Tepic [now the state of Nayarit]."

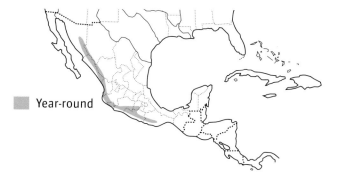

Year-round

Appearance

The Colima pygmy-owl is a tiny owl (length 5–6 inches / 13–15 cm) with short wings, a long tail and no ear tufts. Like other members of this genus it has "false eyes"—two black patches on the nape outlined in white. The head and nape, which are lighter in color than the rest of the body, are grayish tawny brown with extensive whitish to pale buff spotting as well as a narrow cinnamon band below the false eye spots. The owl's actual eyes are yellow, framed by a pale brownish facial disk with short whitish eyebrows. The dorsal plumage is tawny olive brown with pale cinnamon to buffy white spots on the wings and six or seven white to buffy white bars on the tail. The sides of the chest are cinnamon brown and the rest of the underparts are whitish with coarse cinnamon brown streaking. This species' proportionally large, bright yellow-orange feet are a distinctive feature.

Voice

The Colima pygmy-owl's primary song consists of 2 to 24 short, hollow whistled notes separated by long pauses. The half-second pause between the first two notes is longer than the rest; the tempo then becomes fairly steady, though the pauses often become slightly shorter as the song progresses. Singing is typically done in bouts beginning with two- or three-note songs and building to six- to ten-note songs, with an interval of about five seconds between songs. Singing bouts are often preceded by a soft quavering trill. Other vocalizations have not been documented.

Activity Timing and Roost Sites

These owls are active mostly at dusk and dawn but may also be active throughout the day. They are rarely seen or heard at night. No information on roosting is available.

Distribution

The Colima pygmy-owl is found only in western Mexico. Its range covers a narrow strip of land extending from central Sonora to Oaxaca.

Habitat

Colima pygmy-owls occupy a range of habitats from deciduous thorn forest to dry pine/oak woodland, at elevations from near sea level to 4,950 feet (1,500 m). They are found mostly along the foothills of the Pacific slope.

Feeding

The Colima pygmy-owl's diet includes birds, reptiles and insects.

Breeding

Colima pygmy-owls are cavity-nesters and lay two to four eggs in May. Little else is known about their breeding biology.

Migration and Other Movements

The Colima pygmy-owl is nonmigratory. Juvenile dispersal has not been studied.

Conservation

Although there is little quantitative data available to assess this species' status, it is considered common within its range. As a cavity-nester it is vulnerable to habitat changes that reduce the availability of nest sites.

Ferruginous Pygmy-Owl
Glaucidium brasilianum

The ferruginous pygmy-owl was first recorded in the United States in January 1872 by Lieutenant Charles Bendire while he was stationed in Arizona at Fort Lowell, now better known as Tucson. This species was common enough in the mesquite thickets along Rillito Creek, where Bendire collected his first specimen, that he was able to obtain several more that spring and summer. Today there are no ferruginous pygmy-owls living near Rillito Creek, or almost anywhere else in Arizona. While various factors may have contributed to this situation, habitat loss appears to be largely to blame. Historically in Arizona, ferruginous pygmy-owls nested mostly in lush mesquite woodlands and cottonwood/mesquite forests along the Gila, Salt and Santa Cruz rivers and their major tributaries. By 1900 much of the original riparian vegetation in the Tucson area had been cleared by woodcutters, and hydrological engineering was starting to impose even greater changes on the landscape.

In the late 1800s cottonwoods declined along the banks of the Santa Cruz as the river became entrenched, partly as a result of water diversion. While this was bad news for species such as the ferruginous pygmy-owl, the diversion of water from the Gila River and its tributaries into unlined irrigation ditches created a network of new, artificial riparian habitats, where some ferruginous pygmy-owls took up residence. Then in 1902 the National Irrigation Act was passed and the era of large-scale dam-building began. By the 1930s, reduced water flow below the dams, along with other channel changes and destruction of riparian vegetation, had turned many areas that were once prime riverside habitat for ferruginous pygmy-owls into large sandy washes.

Since 2000 Arizona's ferruginous pygmy-owl population has stood at fewer than 40 adults. With at least 85 percent of the state's original riverside forests eliminated, these owls are now found mainly in Sonoran desert scrub and semi-desert grassland habitats. In 1997 the ferruginous pygmy-owl was federally listed as an endangered species in Arizona. However, that designation was rescinded in 2006, despite compelling evidence that populations of this species living in Arizona, Sonora and Sinaloa are genetically distinct from those in Texas and in the rest of Mexico.

Year-round

The eyes are yellow, the bill and feet are light greenish to grayish yellow, and the legs are dull white to pale brown.

Voice

The territorial advertisement call consists of a series of 10 to 30 whistled toots with an upward inflection, delivered at a rate of two to three notes per second. There is a 5- to 10-second pause between each repetition of the song. This vocalization is given mainly by males; they also deliver a quieter version as a contact call when arriving at the nest. Before incubation starts, males may sing continuously for up to five hours. The most common female vocalizations are the chittering food-solicitation call and the alarm call, which consists of two short, sharp, upwardly inflected notes (*pee weeet*) repeated at irregular intervals.

Appearance

The ferruginous pygmy-owl is distinguished by its small size (length 5.5–7 inches / 14–18 cm), short wings, long tail, lack of ear tufts and false eyes (two elongated black patches on the back of the head, bordered with white). The background color of the head and back plumage is generally either grayish brown or rufous, but sometimes intermediate between the two. Characteristic markings include numerous fine white or buff streaks on the head, bold white spots on the shoulders and coarse dark streaking on the whitish underparts. The tail is usually marked with narrow dark and pale bars of color varying from dark or grayish brown and white to dark brown and rufous. In some cases the tail is cinnamon-rufous with only faint barring. The facial disk is dull white with brown to rufous streaking and white eyebrows.

Activity Timing and Roost Sites

Ferruginous pygmy-owls do most of their hunting and vocalizing around dusk and dawn but are also active during the day and night. Those that live in arid scrubby habitats tend to be more active during daylight hours and less shy than those that live in humid rainforest. Roost sites are typically low in the tree canopy.

Distribution

The ferruginous pygmy-owl is found throughout much of South America and discontinuously from Central America to the extreme southern United States. Its North American range extends from southern Arizona and central Sonora south along the Pacific slope, and from southeastern Texas south along the

Gulf Coast. Recent genetic studies suggest that North American ferruginous pygmy-owls should be recognized as a separate species.

Habitat

Ferruginous pygmy-owls occupy a wide variety of habitats across their range, from semiarid desert scrub to tropical rainforest. They are resident from sea level to 4,620 feet (1,400 m) in the United States and Mexico. They are found in riparian mesquite and cottonwood/mesquite woodlands and saguaro desert scrub habitats in Arizona, and in mesquite brush, ebony and riparian habitats of the lower Rio Grande Valley in Texas. In Mexico they are commonly associated with saguaro desert scrub, thorn scrub and tropical deciduous forest in Sonora and with riparian forests along the Atlantic and Pacific coasts.

Feeding

The ferruginous pygmy-owl's diet is dominated in some regions by insects, especially grasshoppers and crickets, and in other areas by reptiles such as lizards and skinks. These owls also eat small birds and mammals up to the size of quails, and cotton rats and the occasional amphibian. They hunt mostly from perches, capturing birds low in the canopy and other prey on the ground or in low vegetation. They also take birds from tree cavities.

Breeding

Males defend territories year-round and may mate with the same female year after year. Ferruginous pygmy-owls usually nest in cavities in trees or cacti—either excavated by gila woodpeckers or gilded flickers or formed naturally—but sometimes nest in tree forks or depressions. They will use nest boxes. The clutch size is usually three to five but ranges from two to seven. Laying occurs between late March and mid-June, mostly from late April to mid-May. Incubation takes 23 to 28 days. The young climb or jump from the nest when three to four weeks old and make their first flights about four days later. Fledglings remain close together in the low canopy or dense understory for about five weeks and are fed by both parents for about eight weeks. The young then disperse abruptly. Limited research indicates dispersal distances of about 5 miles (8 km).

Migration and Other Movements

Ferruginous pygmy-owls are considered nonmigratory. However, it is unclear whether or not some members of breeding populations in Sonora, and perhaps other northern parts of the range, move away from their breeding areas in winter.

Conservation

Ferruginous pygmy-owls are common throughout much of their Mexican range, but populations in northeastern Mexico and northern Sonora have decreased notably in recent times. Populations in Texas and Arizona have declined severely over the past century. Threats include the loss of riparian habitats to livestock grazing, water withdrawal and development, and the loss of desert scrub and semi-desert grasslands to urban expansion and nonnative vegetation introduced to support livestock grazing.

Northern Pygmy-Owl
Glaucidium gnoma

Many pygmy-owls, including all of those found in the Americas, are distinguished by the conspicuous eyespots on their nape—large black oval or teardrop-shaped patches outlined in white—that give the appearance of backward-facing eyes. Common wisdom has long held that these "false eyes" help deter attacks from behind by either predators or mobbing birds, a view neatly expressed by naturalist Charles W. Michael in Arthur Cleveland Bent's *Life Histories of North American Birds of Prey.* "When a pygmy [owl] is discovered all the small birds of the neighborhood band together and do their combined best to make his life miserable," wrote Michael.

"They curse and revile him, but do not dare to strike: a bird with two faces, four eyes, and a fighting heart is a little too tough to tackle."

In 1999 a group of Montana biologists put this idea to the test, at least in terms of the effect on mobbers. The researchers, led by Caroline Deppe, tested mobbing reactions using painted balsa wood models of northern pygmy-owls set on poles in forest clearings. Half of their trials used a model with eyespots and the other half used one without. A tape recording of northern pygmy-owl calls, as well as mountain chickadee and red-breasted nuthatch mobbing calls, helped set the stage. More than 31 species mobbed the wooden owls over the course of the experiment, with nuthatches, chickadees, juncos, warblers and hummingbirds joining in most often. Whether or not the model had eyespots had no effect on how long the mobbing sessions lasted or how many individuals participated, but it did influence how the mobbers behaved around the owl. When mobbing the no-eyespot model, the birds made as many close passes behind the head as in front. With the eyespot model they avoided the false eyes, which seem to represent a "supernormal stimulus" that is scarier than the real eyes.

Deppe and her colleagues suggest two advantages that pygmy-owls may gain from redirecting mobbers into their line of sight: a reduced risk of being injured or killed by an attack from behind, and increased hunting opportunities for owls that decide to make a meal of one of their tormentors. However, they note

■ Year-round

that a more complete understanding of the benefits will require further study.

Appearance

This is a small owl (length 6–7 inches /16–18 cm) with a plump body, short wings, a long tail and no ear tufts. The base plumage color is gray-brown, and varies geographically—palest and grayest in the Rocky Mountains, darkest brown along the Pacific coast, and rufous in southern Mexico and Central America. Against this background the head, throat, back and wings are flecked with whitish spots and the tail is crossed with five or six narrow white bars. The eyespots on the nape are very evident. The true eyes are deep yellow and framed by an indistinct facial disk of whitish and brown bands, with a white chin and eyebrows. The bill is pale yellow. The underparts are mostly white with brownish streaks and the fully feathered legs are dark brown. Feathering is sparse on the feet, which are yellow.

Voice

The primary song is a series of low, hollow-sounding toots repeated for up to several minutes. It is often preceded by a brief higher-pitched trill. There are at least four distinct geographic variations of this song, depending on whether the notes are given singly (as they are north of southern Arizona) or paired (as they are from southern Arizona to Oaxaca) and depending on the tempo (slow along the Pacific coast and fast in interior regions). The primary song is sung by both sexes but more commonly by the male. The northern pygmy-owl's close-contact call resembles a junco's twitter or a cricket's call.

Activity Timing and Roost Sites

Northern pygmy-owls are primarily active and do most of their singing in daytime, particularly around dusk and dawn. Biologists are uncertain whether they hunt at night. They often perch conspicuously on poles, trees or power lines. They roost on tree branches, sleeping with their eyes closed and their body plumage either fluffed up or sleeked down, depending on the weather. Conifers seem to be preferred for roosting. Adults usually roost alone. Recently fledged young sometimes roost together.

Distribution

This species' range extends from southeastern Alaska to Honduras. In Canada and the United States northern pygmy-owls are found discontinuously from the Pacific coast to the eastern slope of the Rocky Mountains. In Mexico they are found in the mountains of southeastern Baja California and the interior ranges from Chihuahua and Coahuila to Oaxaca, as well as in Chiapas.

Habitat

Northern pygmy-owls inhabit a wide range of deciduous, coniferous and mixed forests, often at high elevations—up to 9,900 feet (3,000 m) in the contiguous United States and 12,210 feet (3,700 m) in Mexico—though they nest at elevations as low as 330 feet (100 m) in Alaska. In winter they are sometimes found in other habitats, including yards with bird feeders, where they prey on small birds.

Feeding

Northern pygmy-owls eat a wide variety of prey, concentrating mainly on small birds—up to the size of jays and thrushes—and mammals, including rodents, squirrels and pocket gophers. They sometimes kill prey that are more than twice their size, such as quails and red squirrels. They also eat various insects and, to a lesser extent, lizards and other reptiles. They generally hunt from a perch but sometimes take birds from open or cavity nests.

Breeding

Northern pygmy-owls are cavity-nesters, using mainly abandoned woodpecker cavities. The clutch size ranges from two to seven, and egg laying begins in April. The incubation period is about 28 days and the young fledge when about 23 days old. Fledglings remain close to each other and are fed by their parents for an undetermined length of time. Juvenile dispersal remains unstudied.

Migration and Other Movements

These owls commonly move from their breeding grounds to lower elevations in winter, at least in some parts of their range, but details of these seasonal movements are not known.

Conservation

Because of lack of data, population trends for this species are not known. Local numbers are likely to be reduced in areas where forestry practices limit the availability of nest cavities. When nest boxes have been provided as alternatives to woodpecker cavities, these owls have not used them.

Eastern Screech-Owl *Megascops asio*

It's not uncommon for owls to eat snakes, but when it comes to Texas blind snakes, eastern screech-owls have come up with a better idea. Instead of killing them, they recruit them as live-in janitors.

The Texas blind snake is a 5- to 8-inch-long (13–20 cm) sightless and almost toothless snake that spends its days hiding under logs and rocks. At night it prowls on the surface, hunting for earthworms, insect larvae, ant pupae and termites. When biologists F.R. Gehlbach and R.S. Baldridge found live blind snakes in the debris at the bottom of eastern screech-owl nest cavities at the end of the nesting season, they wondered how they had got there.

After seeing several adult screech-owls arrive at their nests carrying obviously living blind snakes (the reptile's body was coiled around the owl's bill instead of dangling limply), the question became not "how?" but "why?"

Further observations and experiments eventually led Gehlbach and Baldridge to conclude that the owls benefit from bringing blind snakes to their nests and allowing them to burrow out of reach. Eastern screech-owls are poor housekeepers, and their nests become increasingly fouled with uneaten food, fecal matter, pellets and the occasional dead owlet as the nesting season progresses. The soft-bodied larvae of insects that are attracted to this mess are ideal prey for blind snakes.

Gehlbach and Baldridge found that eastern screech-owls raised in nests with live blind snakes grew significantly faster than those in snakeless nests. This suggests that resident snakes enhance nestling growth by decreasing larval parasitism on nestlings or reducing competition between nestlings and insects for cached food. Or perhaps the snakes' consumption of insect larvae is beneficial in both ways.

Elf owls and whiskered screech-owls occasionally eat blind snakes, but no one is sure whether they ever bring these prey to their nests while still alive. In another example of apparent symbiosis involving cavity-nesting owls, tree ants sometimes take up residence with whiskered screech-owls, presumably because of the food scraps found in their nests. The ants don't bother their hosts but will attack intruders.

■ Year-round
--- Rare and local

Appearance

This small owl (length 6–10 inches / 16–25 cm) has conspicuous ear tufts, yellow eyes and an olive yellow bill. Eastern screech-owls occur in two distinctly different color morphs. Gray-morph birds have pale gray underparts marked with dark vertical streaks and bars and gray-brown dorsal plumage streaked with black. Two or three rows of white spots extend down each shoulder. The lightly banded pale gray facial disk is surrounded by a narrow black border. The feathered legs and feet are light gray, buff or white. In the rufous morph the background color of the plumage is pale rusty to bright orange-red and the markings are finer and fainter.

Voice

All vocalizations are given by both sexes. Two trilled songs are sung separately or together along with various calls. The male's nest-site advertising and courtship song, also used by both sexes for pair and family contact, is mostly monotonic and lasts three to six seconds. The territorial defense song is a quavering descending whinny lasting one-half to two seconds. Alarm calls include, in order of increasing anxiety level, soft, low-pitched hoots, usually two to four at a time; loud single high-pitched barks; and loud single piercing or grating screeches.

Activity Timing and Roost Sites

Eastern screech-owls are mainly nocturnal but sometimes begin hunting before sunset. In fall, winter and early spring they roost in tree cavities, nest boxes or conifers. In summer they roost in dense deciduous tree foliage or vine tangles or on a tree limb next to the trunk.

Distribution

This species is resident in most of the United States east of the Rocky Mountains. Its range extends north into extreme southern parts of Canada, from Saskatchewan to Quebec, and south into northeastern Mexico on the Atlantic slope from Coahuila to southern Tamaulipas.

Habitat

Eastern screech-owls inhabit a wide range of coniferous and deciduous forest habitats, mostly below 4,950 feet (1,500 m), including those in urban and suburban areas. Trees with nest cavities are a key element, and an open sub-canopy space with few shrubs is preferred for hunting.

Feeding

Eastern screech-owls eat at least 138 different vertebrate species, mostly songbirds and rodents, and a similar variety of invertebrates, including insects, crayfish and earthworms. They eat small prey immediately but may cache larger prey, usually in a tree cavity. They hunt from perches below the tree canopy, usually making a direct strike from the perch, but sometimes hovering. The average perch height is 8.6 feet (2.6 m).

Breeding

In years of high population density and plentiful food, a small proportion of males mate with two females simultaneously or sequentially. Otherwise, eastern screech-owls are monogamous and usually mate for life. Pairs will make up to four renesting attempts to replace clutches lost to predators or abandoned

because of disturbance. They nest frequently in natural tree cavities and somewhat less often in cavities made by northern flickers or other woodpeckers. They also use a wide variety of artificial sites, including nest boxes, mailboxes and dovecotes. The clutch size ranges from two to six, usually three or four, and laying begins in late March. The eggs are incubated for about a month. The nestling period is about 28 days and fledglings stay with parents for eight to ten weeks. Eastern and western screech-owls sometimes hybridize.

Migration and Other Movements

Eastern screech-owls are nonmigratory. Dispersing juveniles usually settle within 10 miles (17 km) of their natal territory but move farther in severe winters or when food is scarce. Such conditions may also cause adults, especially females, to leave their home ranges.

Conservation

Suspected declines in the past may reflect regular population cycles, as documented in central Texas. Eastern screech-owls seem adaptable to habitat change, provided tree density is not reduced below 50 trees per 2.5 acres (1 ha) and there are sufficient woodpecker cavities and other nest sites. They readily habituate to humans and are often the most common birds of prey in treed suburban and urban areas.

Vermiculated Screech-Owl
Megascops guatemalae

The vermiculated screech-owl's unusual name comes from the Latin *vermiculatus,* meaning "full of worms" or "worm-eaten." However, these owls are neither notable consumers of earthworms nor shabby looking. In ornithological jargon the term vermiculated is used to describe plumage patterned with fine, irregular or wavy lines, like the meandering tracings of worms in wet mud. Such markings—referred to as vermiculations—are common among the genus *Megascops,* but only one species gets to wear this label. Unfortunately the experts have not yet reached a consensus on which owl is the true vermiculated screech-owl.

Some authorities refer to *Megascops guatemalae* as the Guatemalan or Middle American screech-owl and regard it as distinct from *Megascops vermiculatus,* which lives in Costa Rica and farther south. The American Ornithologists' Union considers the two types similar enough to merge into a single species, which they designate *Megascops guatemalae* or, in English, the vermiculated screech-owl. Exactly which South and Central American subspecies come under this umbrella and which rate full species status is still a matter of debate.

DNA analysis suggests that *Megascops guatemalae* is more closely related to the eastern and western screech-owls than to South American screech-owls, but relationships among the members of this genus remain unclear. Meanwhile there is much to learn about the habits and ecological roles of these little-studied owls, regardless of how they are named.

Appearance

This small owl (length 8–9 inches/20–23 cm) has short ear tufts, yellow eyes, a greenish gray bill, feathered legs and bare, brownish pink toes. It occurs in two distinct color morphs. Brown-morph birds have dark gray-brown to black-brown dorsal plumage and lighter underparts, both marked with darker vermiculations and streaks. The facial disk is light brown, accented by a narrow dark border; thin white eyebrows contrast with dark ear tufts. The crown is heavily spotted and barred black. Rufous-morph birds have reddish coloring and are less boldly patterned.

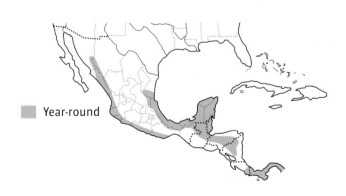

Year-round

Voice

The vermiculated owl's primary song is a monotonic toad-like trill that lasts from 3 to 19 seconds. It begins softly, then gets louder before its abrupt conclusion. It may be repeated a few times in quick succession, with fairly long pauses between sequences. Pairs often perform duets.

Activity Timing and Roost Sites

Vermiculated screech-owls are nocturnally active and seldom seen by day. They roost in dense foliage or cavities.

Distribution

In Mexico, vermiculated screech-owls are found along the Pacific slope from southern Sonora to Oaxaca and along the Atlantic slope from southern Tamaulipas through the Yucatán Peninsula. Their range continues through Central America and parts of South America.

Habitat

Vermiculated screech-owls favor densely vegetated habitats, including evergreen to semideciduous forests, thorn forests, thick scrub, second-growth woodlands and tree plantations. In Mexico they are found from sea level to 4,950 feet (1,500 m).

Feeding

Vermiculated screech-owls eat mainly large insects such as beetles, katydids and grasshoppers. Their diet also includes frogs, reptiles, fish and rodents. They hunt from perches along forest edges or in clearings and capture their prey on the ground, on branches or in the air.

Breeding

Most aspects of the vermiculated screech-owl's breeding biology have not been studied. The clutch size ranges from two to five and the eggs are laid in March to April. The usual nest site is an old woodpecker cavity or natural tree cavity.

Migration and Other Movements

Vermiculated screech-owls are nonmigratory.

Conservation

Within its Mexican range this species appears to be fairly common, but populations have probably declined in response to habitat loss.

Western Screech-Owl
Megascops kennicottii

Throughout their range western screech-owls vary significantly in body size, coloration and foot feathering. Although the experts are still debating how many subspecies should be recognized to account for this complexity— proposals range from 9 to 18—the patterns of variation are quite clear, and they invariably conform to one of the "rules" devised by biologists to explain physical differences between and within species.

The size of western screech-owls increases gradually from south to north and from low-lying coastal regions to higher inland elevations. The smallest individuals live in southern Baja California. The largest are found in the northern interior parts of the species' range, from northern Nevada and northeastern California to southern British Columbia. This trend follows Bergmann's Rule, which

states that individuals living in colder regions generally have larger bodies than their warm-region relatives, because a larger mass results in a lower surface-to-volume ratio and reduces relative heat loss—an advantage in chilly climates and a disadvantage in sweltering ones.

Kelso's Rule, which is based on a study of New World owls, makes several predictions about foot feathering: that bare-toed owls are most frequently found in warm, humid areas; that species with bristles or sparse feathering on their feet and toes are typically associated with arid environments (except in the coldest zones); and that owls living in colder areas have denser and longer foot feathering than those in warmer areas. Western screech-owls fit this model perfectly. Those resident in desert habitats in northwestern Mexico have bristle-covered feet, whereas feathered feet are standard elsewhere. Foot feathering is heaviest in northern populations.

The third ecogeographical principle demonstrated by this species is Gloger's Rule, which asserts that birds (and mammals) of warm, humid regions have darker coloring than representatives of the same species living in cooler or drier areas. Desert-dwelling western screech-owls tend to have pale gray plumage, while rainforest residents in the Pacific Northwest states, British Columbia and Alaska are the darkest members of the species, often with rich red-brown plumage.

Appearance

The western screech-owl is a small owl (length 7.5–10 inches/19–25 cm) with yellow eyes and prominent ear tufts. Base plumage coloring varies: gray-brown to rufous brown in the northern part of the range and along the California coast; pale gray through the American Southwest;

■ Year-round

gray with a buff or tawny wash in southern Baja California and northwestern Mexico; and dark gray in northern Baja California and the Mexican interior. In all regions both the lighter underparts and darker dorsal plumage are patterned with dark vertical streaks and crossbars; these markings vary geographically in coarseness and density. A row of pale spots runs down each shoulder. A narrow dark line borders the generally pale gray facial disk and the bill is either dark gray (in the southern part of the range) or light gray (in the north). The legs are feathered and the feet vary from heavily feathered in the north to bristle-covered in arid parts of northwestern Mexico.

Voice

This species has two common vocalizations, which are given by both sexes. During courtship males and females perform duets as they approach each other. The territorial advertisement and defense song is a series of mellow whistled notes (5 to 15, depending on the subspecies) with an accelerating bouncing-ball rhythm. When a female approaches a singing male during courtship he often switches to a continuous series of these notes that doesn't speed up. The second common call, used for contact between mates, consists of two short trills separated by a momentary pause. The female's food-solicitation call is a descending three-note whinny. The alarm call resembles a coyote's bark.

Activity Timing and Roost Sites

Western screech-owls are mainly active at night but sometimes leave their roosts just before sunset. During the nestling period, parents may hunt by day to keep up with food demands. Winter roosts are typically in cavities, nest boxes or conifers. Summer roosts are usually in deciduous trees, often next to the trunk. Caves and crevices are also used, particularly in hot, arid areas with few trees.

Distribution

The western screech-owl's range extends from southern Alaska to central Mexico. In Alaska and Canada it is confined to the coast and extreme southeastern British Columbia. South of the Canadian border its range widens toward the Great Plains, but with only scattered locations in Oregon, California, Nevada, Utah and Arizona. The Mexican range includes Baja California, Sonora and northern Sinaloa and a wide swath through the interior from eastern Chihuahua and western Coahuila to Puebla. Over the past four decades this species has moved eastward into central Texas. It also seems to be expanding its range east of the Rockies, from southern Alberta to New Mexico.

Habitat

Western screech-owls inhabit a wide array of deciduous woodlands and mixed deciduous and coniferous forests, as well as treed urban habitats. In the Sonoran Desert, densities are highest along stream margins and washes. Elsewhere they favor river-valley bottoms and other riparian habitats. They are restricted to low elevations in the northern part of their range but are found from sea level to 8,250 feet (2,500 m) in Mexico.

Feeding

There is a high degree of geographical and seasonal variation in this opportunistic species' diet. Depending on location, either small mammals or small birds top the list. Other prey include insects (such as moths, caterpillars and crickets), earthworms (brought to the surface by rain), crayfish, frogs and fish. These owls hunt from perches below the tree canopy or projecting just beyond tree foliage. They capture most prey on the ground but also glean insects from foliage, catch insects or bats in flight and pluck fish from shallow water.

Breeding

Western screech-owls defend their territories year-round and often use the same nest site for several consecutive years. Their nests are usually located in northern flicker or pileated woodpecker cavities in trees or gilded flicker cavities in columnar cacti. They use natural tree cavities less often but readily accept nest boxes. The clutch size ranges from two to seven but is usually three to five. Depending on latitude, the eggs are laid between early March and late May. The incubation period is about 26 to 30 days, followed by a nestling period of similar length or slightly longer. Juveniles disperse about two months after leaving the nest. Eastern and western screech-owls sometimes hybridize.

Migration and Other Movements

This species is nonmigratory. Dispersing juveniles probably move about 3 to 9 miles (5–15 km).

Conservation

Population trends are unclear but slow declines are suspected in some areas. Loss of riparian habitat is the greatest threat overall. In the Pacific Northwest, where barred owls are relatively recent arrivals, they may be reducing western screech-owl populations. The western screech-owl's eastward range expansion is likely a response to tree planting on the Great Plains.

Whiskered Screech-Owl
Megascops trichopsis

Few birders, whether they are dedicated life-listers or casual watchers, are immune to the thrill of seeing a bird they have never before encountered. That's one reason why places like Cave Creek Canyon in Arizona's Chiricahua Mountains are so popular. Located just north of the Mexican border, the Chiricahuas are home to many animals and plants that are found nowhere else in the United States, a great attraction for Americans who want to see these species without leaving their own country.

But what's good for nature enthusiasts is not always good for nature. In springtime the number of birders that visit Cave Creek Canyon can exceed two thousand a month. They come to see species such as the elegant trogon, the rose-breasted becard, the blue-throated hummingbird and the whiskered screech-owl, and sometimes their enthusiasm for pursuing these rarities has negative consequences for the objects of their desire. Among the activities that can be detrimental to whiskered screech-owls are the repeated use of tape-recorded or imitated calls to lure owls into view, knocking or rubbing on nest trees to prompt the female to peek out of the nest hole, shining bright lights for nighttime observation, and using camera flashes. One pair of nesting whiskered screech-owls was regularly watched and photographed at dusk for six weeks by up to 14 people a night. They reacted to this attention by initiating evening activity later than normal and switching most of their feeding to dawn. Only two of their three eggs hatched and the two owlets were underweight when they fledged.

The south fork of Cave Creek Canyon is now designated as a special biological area where playback of taped calls is prohibited in the nesting season. Biologists Frederick and Nancy Gehlbach, authors of the whiskered screech-owl account in *The Birds of North America*, say that even whistled imitations of calls should be used sparingly, and they offer additional guidelines for people seeking these and other cavity-nesting owls during nesting. They suggest looking for males on their day roosts, which in the case of whiskered screech-owls are usually fewer than 11 yards (10 m) from the nest, instead of disturbing the nest with noise or lights at night. They also recommend approaching owl-watching "as a chance to exercise [your] senses and enjoy [the owl's] concealing plumage and posture."

■ Year-round

Appearance

At 6 to 7.5 inches (15–19 cm) in length, this is the smallest of the four North American screech-owls. It is named for the whisker-like bristles on the ends of its facial disk feathers, though these are visible only at close range. Whiskered screech-owls have ear tufts, an olive yellow bill and golden to orange-yellow eyes. Their legs are heavily feathered, but the toe feathers may be sparse or bristle-like. Females are darker than males, especially within populations in the United States and northern Mexico.

There are two color morphs. Gray-morph birds have gray base coloring with rusty or, less often, brownish overtones. The lightly banded gray and buff-rust facial disk is highlighted by a black border, white-edged ear tufts, a dark V that extends from bill to forehead and a dark buff to rusty throat patch. The pale buff to light rust underparts are marked with dark streaks and bars. The dorsal plumage is brownish gray, becoming paler around the nape, and is streaked darker brown. When the wings are folded, a broken line of whitish spots may be visible on each side. The rufous morph is less common and occurs mostly from central Mexico south. These birds have pale cinnamon-buff to reddish brown base coloring.

Voice

All whiskered screech-owl vocalizations, except for two variations on the primary song, are given by both sexes. The primary song is a trilled series of 3 to 15 notes. One male-only variation is a short trill that rises, then falls; the other is a long trill that can extend for more than a minute. The secondary song, used mainly for pair coordination, is a syncopated series of 10 to 20 notes in various Morse code–like rhythms. Pair duets are usually antiphonal (with alternating voices) but sometimes synchronized. The contact call is a whistled *kew*. Alarm is signaled by a hoot, bark or screech, depending on the level of agitation.

Activity Timing and Roost Sites

Whiskered screech-owls are primarily nocturnal. They usually roost in coniferous or broadleafed trees, generally in dense foliage or next to the trunk. Cavity roosting is less common.

Distribution

The whiskered screech-owl's range extends from extreme southeastern Arizona and southwestern New Mexico to northern Nicaragua but is restricted by elevation, becoming interrupted by lowlands in central and southern Mexico and Central America. In northern Mexico it follows the Sierra Madre Occidental south from northeastern Sonora and the Sierra Madre Oriental south from southwestern Tamaulipas.

Habitat

This species is a resident of mountain woodlands and forests at elevations of 3,300 to 9,570 feet (1,000–2,900 m). In the United States and northern Mexico its preferred habitats are canyon riparian forest, pine/oak and other types of evergreen woodland, and pine forest. Juveniles that disperse between mountain ranges probably keep to wooded areas in the foothills and valleys.

Feeding

Insects, especially moths, caterpillars and beetles, dominate the diet. Small mammals and reptiles are also important prey. Whiskered screech-owls hunt from perches, making repeated short, direct flights out to seize insects in flight or, less commonly, to capture prey on the ground, in foliage or on a trunk or branch. They occasionally hover above the leaf litter or among branches.

Breeding

Whiskered screech-owls nest in natural tree cavities and those created by northern flickers or other large woodpeckers, or in the decay-hollowed tops of broken trunks. They rarely use nest boxes. In the northern part of the range their two to four eggs are laid between early April and mid-May. Incubation takes about 26 days and the nestling period is 24 to 30 days. The young remain with their parents for at least a month after fledging.

Migration and Other Movements

Whiskered screech-owls are nonmigratory. In southeastern Arizona, and perhaps elsewhere, individuals living at higher elevations may move downslope in winter. Juvenile dispersal distances are not known.

Conservation

This species is classified as threatened in New Mexico because of its restricted range. Elsewhere it is common to fairly common, but vulnerable to habitat change, especially loss of nest trees, resulting from logging, fire suppression or overgrazing.

Elf Owl *Micrathene whitneyi*

Early North American ornithologists considered the elf owl strictly a desert resident that invariably nested in abandoned gila woodpecker or gilded flicker cavities in saguaro cacti. That view changed in the 1930s, when two independent naturalists published articles about their bird studies in southern Arizona. Berry Campbell was first off the mark, describing his discovery that elf owls not only nested in the wooded foothills around Pena Blanca Spring but were also the most abundant owl in the area. "It has generally been held that to look for these birds out of the sahuaro cactus belt was a waste of time," he wrote. "Surely we must revise our ideas concerning the habits of the bird."

That view was backed up a few years later when Herbert Brandt detailed his observations from a visit to the western slopes of the Huachuca Mountains. "I was surprised to find the Elf Owl not uncommon here," he reported, "whereas this strange little creature is supposed to stray but seldom from the arms of the giant cactus on the lower desert. A pair of these little birds occupied a Flicker's hole in a large tree…and each evening after the blanket of night was spread, the little male perched near the top of a medium-sized tree, where he whimpered and whistled continuously for some time."

Despite this long-held belief in the elf owl's desert affinities, this species may be a latecomer to such arid habitats. It is only during the past eight thousand years that the present-day Sonoran region has been dominated by saguaros and associated desert plants. Earlier it was occupied by evergreen woodlands and riparian forests, and it is in this environment that the elf owl likely evolved. The presence of elf owl bones dating back at least 11,000 years within caves in the region supports this premise. Other evidence that elf owls may be secondary desert inhabitants includes physiological research showing that they find it difficult to avoid overheating when humidity is low and cannot survive prolonged body temperatures above 108°F (42°C)—unlike whiskered screech-owls, which can endure such high body temperatures with no ill effects. Equally telling is the fact that elf owl populations are denser and more stable in subtropical thorn woodlands, montane evergreen woodlands and riparian forests than in saguaro cactus country.

Breeding
Year-round
Wintering

Appearance

The elf owl has a short tail and no ear tufts, and at 5 to 5.5 inches (12–14 cm) in length it is the world's smallest owl. It has yellow eyes and prominent white eyebrows. The facial disk is cinnamon-buff with white spots along the lower edge, and the bill is greenish yellow. The gray-brown dorsal plumage is mottled buff to cinnamon and marked with a narrow white nape band, an irregular white stripe down each shoulder, and white spots along the outer edges of the folded wings. Brownish to cinnamon streaks mark the white underparts. The unfeathered legs and feet are dull yellow.

Voice

The male elf owl's distinctive primary song is a series of high-pitched puppy-like yips. It usually consists of five to seven notes and lasts about a second, but it is faster and longer when given from a potential nest cavity. The main female-only call is a cricket-like trill uttered when receiving food from her mate. Both sexes give a single short, slurred whistle as a contact call and single or rapidly repeated sharp barks to signal alarm or when mobbing potential predators.

Activity Timing and Roost Sites

Elf owls are strictly nocturnal. They often roost in cavities in trees and columnar cacti, in mistletoe clumps and in evergreen trees, especially alligator junipers. Roosting in deciduous tree foliage is less common.

Distribution

Three separate breeding populations span the U.S.–Mexico border between eastern California and the Gulf of Mexico. Three other, exclusively Mexican populations live in southern Baja California and Puebla. The borderland populations winter in Mexico.

Habitat

Elf owls inhabit a variety of arid to semiarid habitats at elevations from 330 to 6,600 feet (100–2,000 m). These include desert, desert-wash woodland, subtropical thorn woodland, canyon riparian forest and pine/oak woodland.

Feeding

Elf owls eat mainly insects, especially moths, beetles and crickets. They occasionally eat lizards, blind snakes and young kangaroo rats. They hunt mostly from perches, darting out or swooping down to catch prey in flight, on the ground or on branches. Less common hunting techniques include running after prey on the ground, flailing insects from vegetation and walking on or hanging from flowers or foliage while probing for insects.

Breeding

Elf owls are usually monogamous, but one case of polygyny is known. For nesting they depend on cavities made by small to medium-sized woodpeckers in columnar cacti, deciduous trees and large yucca and agave flower stalks. They will use nest boxes. The clutch size ranges from one to five. Egg laying begins in late April or early May in the southern United States and northern Mexico. Incubation takes about 21 to 24 days and the young fledge about a month after hatching. They can catch some live prey almost immediately but are fed by their parents for an unknown length of time.

Migration and Other Movements

The borderland populations are migratory, though a few individuals may remain during warm winters. Depending on location and elevation, the migrants return to breeding areas in mid-March to mid-April and leave by mid-October.

Conservation

As a result of habitat loss this species has been extirpated from northern Baja California, is almost gone from California and is declining in parts of Arizona. Elf owls were not seen in the lower Rio Grande Valley from 1894 to 1959. They have since returned but remain vulnerable to loss of riparian habitat. Throughout their range they are dependent on woodpecker cavities.

Flammulated Owl *Otus flammeolus*

Since 1891 ornithologists have been arguing about whether flammulated owls dine exclusively on invertebrate prey or whether they kill vertebrates now and then, like all the other insectivorous owls in North America. Various authorities reported that flammulated owls did sometimes eat small mammals and birds, but their accounts were invariably based on circumstantial evidence, which some critics dismissed as too flimsy to support claims that the owls actually dispatched these prey themselves.

Recently biologists M. David Oleyar, Carl D. Marti and Markus Mika played detective to try to determine conclusively whether flammulated owls ever kill vertebrates or if the reports linking them to vertebrate victims were merely cases of scavenging, mistaken identity or even a different species of owl delivering food to nestlings that weren't its own. The trio came up with several "smoking guns" but no witnesses, so the jury is still out. Nevertheless, their review of past testimony and summary of their own observations are a reminder that many ornithological mysteries are still to be solved.

Between 1994 and 2001 Oleyar, Marti and Mika found vertebrate remains or other evidence in 10 different flammulated owl nests in two northern Utah populations. Nine of the nests were occupied by 10- to 18-day-old nestlings, while the young had fledged a week earlier from the 10th nest. Altogether they identified 13 individual prey items, including mice, bats and songbirds. One discovery was made when they visited a nest for banding and found a nestling with a mouse's hind feet and tail hanging out of its mouth. Fortunately they got a good look before the owlet swallowed the evidence. Using a video camera mounted inside a nest box, Oleyar also documented a male delivering a mouse to his brooding mate. The tape showed the female toying with the carcass, but the recorder batteries died before the camera captured any images of her eating it or feeding it to her young, though she probably did one of the two, since the mouse subsequently disappeared.

As intriguing as these observations are, more than 99.99 percent of the prey consumed by the flammulated owls in these studies were invertebrates—mostly moths and beetles—confirming this species' reputation as a highly specialized insectivore.

■ Breeding
■ Year-round
■ Nonbreeding

Appearance

The diminutive flammulated owl (length 6–6.8 inches /15–17 cm) is North America's only small owl with dark brown eyes. Its short ear tufts are usually flattened, creating a square-headed appearance. When perched, its relatively long, pointed wings extend slightly beyond its short tail. This species' body plumage is gray with geographically variable black streaks and bars and rusty overtones. Great Basin and Rocky Mountain birds are blackest, with the broadest markings and least rufous coloring; markings become finer and redness increases to the northwest and southeast. The faintly banded gray and white facial disk is also rufous tinted, especially around the eyes, and is defined by a dark rufous to black border and short white eyebrows. When the wings are folded, a tawny to rufous line shows across the shoulders. The legs are feathered and the toes and bill are gray-brown.

Voice

The ventriloquial quality of the flammulated owl's hoots often makes it difficult to locate singing owls. The male's advertising song is a single short, low-pitched *hoo,* sometimes preceded by one or two shorter, lower notes. Early in the breeding season it may be repeated for hours. Later, mated males sing more sporadically and mostly late at night, while unmated males continue incessant nocturnal hooting. The female's similar hoots are longer, higher-pitched and more quavering. Both sexes also utter quiet hoots when communicating with mates or young, and escalating barks, moans and shrieks when predators approach the nest or fledglings.

Activity Timing and Roost Sites

Flammulated owls are strictly nocturnal. They roost on tree branches next to the trunk, often in dense vegetation.

Distribution

This species has a patchy distribution because suitable habitat is discontinuous and geographically restricted. The breeding range extends from south-central British Columbia through the western United States (east to Wyoming, Colorado and New Mexico) and down the Sierra Madre Occidental and Oriental to central Mexico. Flammulated owls winter from central Mexico to El Salvador.

Habitat

Flammulated owls nest in cool, dry open forests, often with ponderosa pine as the dominant tree. Large old trees, scattered thickets and clearings are key elements. Little is known about their winter habitat. At the northern end of their range they breed at elevations from 1,320 to 4,455 feet (400–1,350 m). In Mexico they are found from 4,950 to 9,900 feet (1,500–3,000 m). During migration they may pass through lowland areas.

Feeding

Flammulated owls feed almost exclusively on nocturnal insects, especially owlet moths, beetles, crickets and grasshoppers. They hunt from perches and seize their prey in the air, on the ground or from tree trunks or foliage.

Breeding

Flammulated owls mate for life. If both members of a pair return in spring, they claim the previous year's territory. If only the male returns he occupies his old territory with a new mate, whereas a widowed female will find a new mate in an adjacent territory. Preferred nest sites are cavities made by large to medium-sized woodpeckers, but these owls sometimes use natural tree cavities and, occasionally, nest boxes. The clutch size ranges from two to four. Depending on latitude, the eggs are laid from mid-April (northern Mexico) to early June (British Columbia) and incubated for 21 to 24 days. The young fledge at about 23 days old and become independent of their parents after 25 to 32 days, dispersing a few days later.

Migration and Other Movements

Flammulated owls that breed in Canada, the United States and probably northern Mexico are migratory. They travel at night, moving south to southern Mexico and Central America from August to November and returning in late April to early May.

Conservation

Although this is probably the most common owl of montane pine forests in the western United States and Mexico, population trends are undocumented, and a low reproductive rate makes this species sensitive to change. In British Columbia the flammulated owl is listed as a species of special concern. Habitat loss due to logging and fire suppression is the main threat in all areas.

Barred Owl *Strix varia*

Historically the barred owl's range in Canada and the United States was limited to areas east of the Great Plains and south of the boreal forest. Starting in the early 1900s this species began a westward expansion, moving across wooded regions of the Prairie provinces, then north into the Yukon and southeastern Alaska and south through Idaho and the Pacific Northwest to central California. Barred owls were first recorded in Alberta in 1932, in British Columbia in 1943, in the northwestern states in the late 1960s and early '70s, and in California in 1981.

Some authorities suggest that this rapid colonization of the West was promoted by the planting of shelterbelts and establishment of riparian forests by prairie settlers and by logging from the Rockies to the Pacific coast. Others contend that it was an inevitable natural development, perhaps linked to climatic changes. Regardless of why it happened, the recent arrival of the barred owl has had a definite impact on some of its new neighbors, particularly the closely related spotted owl.

The barred owl's range now overlaps the entire range of the northern spotted owl, a subspecies that is federally listed as threatened in the United States and is effectively extirpated in Canada. This is bad news for the spotted owls, since their slightly larger and more aggressive cousins almost invariably win contests for food and territory. Not only do barred owls tend to displace spotted owls, they also sometimes kill them. And because barred owls are more flexible in their habitat requirements, occupying anything from undisturbed old-growth forest to treed urban areas, they have a competitive edge in regions where logging or development is widespread.

Concerns about the effects of interbreeding have also been raised, though hybridization appears to be a relatively rare event so far. Some biologists say it is unlikely to pose a serious threat to either species, and may become even rarer as barred owls become more numerous and can more easily find mates of their own kind. The opposing view is that even low rates of interbreeding could affect the northern spotted owl both directly, by influencing reproductive success and population dynamics, and indirectly, by jeopardizing its conservation status, since the U.S. Endangered Species Act does not protect hybrids.

Year-round

Appearance

Both sexes of this medium-large owl are similar in length (17–20 inches/43–50 cm), but females are noticeably heavier. The barred owl's distinguishing features include an absence of ear tufts, dark brown eyes, yellow bill and extensively barred plumage. The head, neck, upper breast and back are pale buff with dark brown barring. In contrast with the barred collar, the whitish lower breast and abdomen are marked with long, bold dark brown streaks. The tail is broadly barred dark and light brown. The wings are dark brown with white spotting and barring. A dark border frames the grayish facial disk and fine brown barring creates concentric semicircles around the eyes. The legs and feet of most barred owls are feathered (buff to white, with brown spotting or barring), though the feathers may be sparse or bristle-like near the ends of the toes. The southeastern U.S. subspecies has bare, dull yellow toes.

Voice

This is one of North America's most vocal owls, and all calls are given by both sexes. The most familiar call is popularly rendered as "Who cooks for you? Who cooks for you all?" However, the phrases sometimes have fewer syllables. Each phrase begins with two or three distinct, low-toned *hoo*s and finishes with a loud, emphatic note that runs into a final *hoo-ooo*. Another common call consists of six to nine regularly spaced ascending hoots followed by a downwardly inflected *hoo-aw*. Both calls likely have territorial and mate-contact functions. Mates sometimes call together, producing a raucous jumble of cackles, hoots, caws and gurgles that has been compared to maniacal laughter. Potential nest predators may elicit strident screams.

Activity Timing and Roost Sites

Although mostly nocturnal, barred owls sometimes hunt during the day. Adults roost on branches in dense foliage (in either coniferous or deciduous trees) or in tree cavities. Recently fledged juveniles sometimes roost on the ground in tall grass.

Distribution

The barred owl's present range covers most of southern Canada and the United States east of the Great Plains; it also extends west along the southern edge of the boreal forest and widens to encompass much of the West, from southern Yukon to central California. Isolated Mexican populations are found along the Pacific slope and in the adjacent interior from northern Durango to Guerrero, and along the Atlantic slope and interior regions from Veracruz to Oaxaca.

Habitat

Barred owls are forest-dwellers, typically in mixed deciduous/coniferous forests, and may have an affinity for sites near water. In parts of the West they also inhabit conifer-dominated forests. They favor large, unbroken expanses of mature or old-growth forest but will use disturbed habitats.

Feeding

This opportunistic hunter eats mammals up to the size of young rabbits and birds up to the size of grouse, as well as amphibians, reptiles, fish and invertebrates. Small mammals make up the greatest portion of the diet. Barred owls hunt mostly from perches and will take prey that is on the ground, on a tree trunk or branch or in water. They occasionally run after terrestrial prey or wade in shallow water when hunting fish or aquatic animals.

Breeding

Barred owls strongly defend their breeding territories year-round and probably mate for life. Nest sites are often reused for many consecutive years, though not always by the same individuals. They nest mainly in natural tree cavities or hollowed-out tops of broken trunks, but sometimes use stick nests made by other birds or squirrels. One ground nest and one nest in an abandoned shed have been found. They readily use nest boxes. The clutch size is one to five, usually two or three. Laying generally begins in March or April, but sometimes in December in the southern United States. The incubation period is 28 to 33 days and the young fledge at four to five weeks old. They start flying at about ten weeks but are not completely independent until four or five months old. When parental feeding stops, the juveniles disperse. Barred and spotted owls sometimes hybridize.

Migration and Other Movements

Barred owls are nonmigratory, with the possible exception of some individuals within the most northerly populations, which may winter south of their breeding areas.

Conservation

Because of their large territories, barred owls occur at low densities. Continent-wide this species' numbers are increasing as its range expands. However, populations are declining in some eastern areas (including the Atlantic provinces, Ontario, Illinois, Iowa and Tennessee) as remaining tracts of extensive mature forests are cleared for timber harvest, agriculture and urban development.

Great Gray Owl *Strix nebulosa*

Nest-box programs, which provide homes for hole-nesting birds, have become a common conservation initiative. But cavity-nesters are not the only owls that can benefit from assistance in a tight real estate market. In forests that are intensively managed for timber production, great gray owl nest sites are typically in short supply. Artificial nest structures can compensate for this lack—as long as the location has other key habitat elements such as hunting perches and leaning trees, which flightless juveniles can climb to safety.

The first biologists to experiment with providing artificial nests for great gray owls spent hours collecting bundles of branches and hauling them up suitable trees to construct imitation hawk nests. Since then researchers have developed a less labor-intensive alternative to these hand-built abodes. The standard wooden nest platform for great gray owls is 24 by 24 inches (60 by 60 cm) in area and 8 inches (20 cm) deep, with angled sides. Drain holes are drilled in the bottom and a thick layer of wood chips is added so the female can create a depression to hold her eggs. Ideally the structure should be mounted at least 50 feet (15 m) above the ground and positioned so that the tree canopy provides shade and hides it from avian predators. The area immediately around and above the platform should be open so that the nesting birds have easy access.

Great gray owls readily accept nest platforms of this design, and a three-year study conducted in Oregon revealed that they had better nesting success with platforms than with natural nest sites. Unfortunately, great horned owls claimed some platforms, depriving the great grays of these sites and potentially increasing the population of the larger species, which preys on fledgling great grays.

Owl enthusiasts who want to provide nesting opportunities for great grays should locate platforms well away from places frequented by humans, since both great gray and great horned owls may defend their nests aggressively. It is also important that there be a dense stand of trees around or near the nest to provide protective cover for the vulnerable juveniles.

■ Year-round
▪▪ Limits of occasional
　　occurrence in winter

Appearance

This owl is recognized by its substantial size (length 24–33.5 inches/61–84 cm) and gray-brown coloring. Other features include a distinct and heavily ringed facial disk, relatively small yellow eyes separated by whitish crescents, white "bow-tie" markings on the neck below a yellow or ivory bill, an absence of ear tufts and a long, wedge-shaped tail. The great gray owl's plumage is a mix of white, gray and browns. The underparts, including the fully feathered legs and feet, are subtly streaked and barred. The dorsal plumage is mostly mottled. Fine gray-brown barring forms six to eight concentric circles on the whitish facial disk, which is framed by a dark brown border. Females are noticeably bulkier than males.

Voice

The great gray owl's most common vocalization, given by both sexes, consists of 8 to 12 low, mellow hoots separated by intervals of about half a second. A softer version may be given by males as a food-delivery call. Repeated soft double hoots are given by both sexes, possibly as a contact and territorial defense call. Females often utter a single or repeated soft *whoop* as a food-demand and contact call. High-pitched chatter accompanies food exchanges between mates or parents and offspring. Alarm is expressed by loud squawks, barks, wails and squeals.

Activity Timing and Roost Sites

Great gray owls regularly hunt during daylight hours, especially around dusk and dawn, as well as at night. Daytime hunting is most common during the nestling period and in winter.

They typically roost on tree branches close to the trunk; during hot weather they choose trees with fairly dense foliage.

Distribution

Great gray owls are found across much of northern North America and Eurasia. In North America their range stretches across the boreal forest region from northern Alaska to southwestern Quebec. It also extends to the Pacific coast in southern Alaska and northern British Columbia and runs south along the Rocky Mountains, the Sierra Nevada and other ranges to east-central California and northwestern Wyoming.

Habitat

Within the boreal zone, great gray owls typically reside in areas of extensive, dense forest interspersed with bogs, meadows and other small openings. They avoid large open areas. In the southern part of their range they breed in deciduous or coniferous montane forests at up to 9,240 feet (2,800 m), moving in winter to lower elevations where the snow cover is thinner. During irruption years, individuals that winter south of the breeding range are sometimes seen in atypical habitats such as open fields or towns.

Feeding

Great gray owls eat mainly small mammals, especially rodents. Voles dominate their diet in much of North America, but pocket gophers are their primary prey at the southern end of the range. They occasionally eat larger mammals (including weasels and snowshoe hares), birds and frogs. These owls hunt mostly from perches but sometimes fly over fields searching for prey. They often hover before striking.

Breeding

Great gray owls are considered monogamous, but occasional polygyny is suspected. In boreal regions pair bonds are not permanent, though mates may reunite in successive years if food is plentiful. Elsewhere they probably mate for life. The most common nest sites are stick nests made by other birds, followed by large broken-topped trees and mistletoe clumps. This species readily uses artificial nesting platforms. Ground nesting occurs very rarely. The clutch of two to five eggs is initiated between March and early May; heavy snow in mountainous southern parts of the range delays laying. The incubation period is about one month. The young leave the nest when three to four weeks old and can fly a week or two later. They become independent by late summer. The usual age of first breeding is three years.

Migration and Other Movements

Although this species is mostly nonmigratory, great gray owls that breed in mountainous areas of the western United States regularly winter at lower elevations. Nomadic wandering in response to prey scarcity occurs in all regions, resulting in irregular irruptions in southern parts of the breeding range and further south, and as far east as coastal Massachusetts. Dispersing juveniles may remain close to home or move hundreds of miles from where they were raised.

Conservation

Population trends have not been determined, but the great gray owl's affinity for mature and old-growth forests makes it vulnerable to habitat changes resulting from commercial timber harvesting and fire suppression. In the boreal forest, peat extraction is also a threat. In southern parts of the range the forest management practice of poisoning pocket gophers in clearcuts is a concern. It is not known whether owls are ever harmed by eating poisoned pocket gophers, but reduced abundance of this important prey species could be detrimental.

Spotted Owl *Strix occidentalis*

Probably no North American bird has been so extensively written about as the spotted owl, particularly the northern subspecies. The recorded history of the spotted owl dates back to 1859, when collector John Xantus announced his discovery of this new species the previous year. Following this debut the spotted owl received an average amount of attention in scientific accounts. It was not singled out for special attention until 1972, when Eric D. Forsman chose it as the subject of his master's thesis research, little knowing that his work would mark the beginning of a long and often acrimonious conflict over forest management in the West.

In 2000, at an international conference on the ecology and conservation of owls, Forsman looked back over his three decades of northern spotted owl research and discussed the unique influence this one bird has had on the public and scientific debate about how American federal forest lands should be managed. Before Forsman and other biologists began their studies in the early 1970s, the spotted owl was considered secretive and rare. Their work revealed that this species was actually widespread and relatively common throughout much of its range, but that it was largely dependent on old forests, which were fast disappearing. Around the same time the environmental movement was gathering momentum and starting to challenge the prevailing wisdom that old-growth forests were stagnant biological wastelands that should be cut down and replaced with carefully managed stands that would be tended, harvested and replanted at regular intervals. The fact that the spotted owl is both photogenic and relatively fearless around humans made it the perfect "poster child" for the environmentalists' cause.

The battle over the fate of the northern spotted owl—and, by extension, the old-growth forests of the Pacific Northwest—raged through the 1980s, both in the courts and on the ground. In 1990 the subspecies was federally listed as threatened under the U.S. Endangered Species Act. Four years later the Forest Service and Bureau of Land Management adopted a regional management plan that restricted old-growth logging, partly in deference to the spotted owl but also in recognition of the needs of a broad range of forest species. Since many people on both sides of the debate remain dissatisfied with this plan, the spotted owl's time in the media spotlight is not likely to end soon.

Northern Spotted Owl

California Spotted Owl

Mexican Spotted Owl

Recent year-round range

Historic range of Northern Spotted Owl

Appearance

This is a medium-large owl (length 18.8–19.2 inches / 47–48 cm) with dark brown eyes, a greenish yellow bill and no ear tufts. Its chestnut to chocolate brown head, neck, back and underparts are dotted with white spots of varying sizes (small on the crown, larger on the back and underparts). The flight feathers of the wings and tail are dark brown with light brown to white barring. The feathered legs and feet are light brown. The facial disk is faintly ringed with light and dark brown, and the eyes are separated by white crescents.

Compared to the other two subspecies, the northern spotted owl is larger and has darker plumage with smaller spots. The Mexican subspecies is the smallest and has the lightest brown plumage and largest spots. The California subspecies is intermediate in size and coloring but more similar to the northern than the Mexican.

Voice

Spotted owls of both sexes share a varied vocal repertoire of low hoots, barks and whistles. The most common vocalization is a territorial call that consists of four high-pitched hoots delivered with an uneven rhythm: *hoo–hoo-hoo–hooo*. With changes in pitch, intensity and number of hoots (up to 15 in a row), this call serves various functions. Rapidly repeated loud, high-pitched barks are commonly given during nest defense. Other vocalizations include a hollow whistle ending with an upward inflection (*cooo-weep*), used for pair contact and food solicitation; a loud grating *wraak* that signals alarm or distress; and rapidly repeated low hoots given during courtship and when delivering food to the young.

Activity Timing and Roost Sites

Spotted owls are mainly nocturnal but sometimes hunt by day, especially during nesting. While roosting during the day they sometimes make opportunistic strikes when prey pass close by. They roost on tree branches, typically in positions that are cool and shaded by a dense tree canopy, often near streams. To avoid heat stress they regularly move short distances to new roosts during the day. They tend to roost higher in the forest canopy and closer to the tree trunk in winter than in summer. They seek shelter from rain under overhanging branches.

Distribution

The northern subspecies inhabits coastal mountain ranges from extreme southwestern British Columbia to California's San Francisco Bay, and interior mountains as far south as Pit River in northern California. The California subspecies is found in the Sierra Nevada and other ranges from south of Pit River to central California and intermittently along the coast from central California to northern Baja California. The Mexican subspecies inhabits scattered mountain ranges and deep canyons across the southwestern United States and through Mexico's interior mountains, from northern Sonora to Michoacán and from central Coahuila to San Luis Potosí.

Habitat

Spotted owls use a wide variety of forest types from sea level to 3,960 feet (1,200 m) in the northern part of their range and at elevations of up to 8,910 feet (2,700 m) in the southwestern United States and Mexico. Northern spotted owls favor relatively dry coniferous forests. California spotted owls use forests dominated by oaks (or other broad-leafed trees) at low elevations, mixed forests at middle elevations and coniferous forests higher up. Mexican spotted owls are found mostly in coniferous and pine/oak forests, but also in steep, narrow canyons that may be either forested or sparsely vegetated. The northern and California subspecies require extensive areas of old-growth or mature forest with a dense, multi-layered canopy. Mexican spotted owls also show a distinct preference for old-growth and structurally complex forests, but have smaller home ranges than the other subspecies.

Feeding

Spotted owls eat primarily small and medium-sized mammals, especially rodents, along with some insects, birds, amphibians and reptiles. Their main prey varies geographically. Northern flying squirrels dominate the northern spotted owl's diet in wetter parts of its range. Bushy-tailed, dusky-footed or Mexican woodrats are more important elsewhere. Spotted owls usually hunt from perches and take prey on the ground or on tree limbs or trunks. They sometimes capture moths or other insects and bats on the wing.

Breeding

Spotted owls generally form long-term pair bonds, with mates occupying a shared home range year-round. Most pairs do not breed every year and some fail to reproduce for up to six years in a row. Nest sites include tree cavities, hollowed tops of broken tree trunks, stick nests made by other birds, squirrel nests, mistletoe brooms and cliff holes and ledges. Cliff nests are rare for the northern and California subspecies but fairly common for the Mexican. Artificial nest cavities are occasionally used. The clutch of one to three eggs (rarely four) is initiated in early March to mid-April and incubated for about one month. Most young leave the nest about 30 to 36 days after hatching, and start flying about a week later. They remain with their parents for two to three months and then disperse.

Migration and Other Movements

Members of the northern subspecies do not migrate. Some California and Mexican spotted owls migrate to lower elevations after breeding, traveling distances of 9 to 40 miles (15–65 km) for an altitude change of 1,650 to 4,950 feet (500–1,500 m). One owl tracked in Utah moved to a higher location for winter.

Conservation

Populations of the Mexican and northern subspecies are declining throughout their ranges, mainly as a result of habitat loss. The California subspecies is generally stable. In the United States the Mexican and northern subspecies are federally listed as threatened. In Canada the spotted owl is designated endangered. In British Columbia, the only province where it occurs, its population has declined precipitously in recent decades and is likely no longer viable.

Northern Hawk Owl *Surnia ulula*

Living far from most humans and scattered across the boreal region in low numbers, the northern hawk owl is one of the least studied birds in Canada and the United States. With its northern haunts under rapidly escalating pressure from industrial activities—logging, oil and gas extraction, mining and construction of hydroelectric dams—the need to know more about this and other boreal forest owls is gaining urgency.

One recent study that provided new insights into the ecology of northern hawk owls examined the relationship between this species and forest fires. Wildfire is a natural process that plays an important role in shaping boreal forest ecosystems. Over the past half-century there has been an increasing emphasis on suppressing forest fires and on salvage logging any stands that do burn. Forest management plans that eliminate older, more fire-prone stands have also changed wildfire dynamics. In the late 1990s biologists Kevin C. Hannah and Jeff S. Hoyt looked at how these activities might affect hawk owls.

Working in conifer-dominated boreal forest in east-central Alberta, Hannah and Hoyt found that these owls were strongly associated with recent wildfires. All nine of the nest sites they located were in areas that had burned within the previous four years, and they saw no hawk owls in nearby mature or old unburned forests. The three nests that they were able to pinpoint were all in large-diameter trees—two trembling aspens and one white spruce—that had been hit hard by the flames. Each tree had a large opening in its trunk where a branch had burned all the way to its base and into the sapwood, creating an ideal nest cavity for a northern hawk owl.

Recently burned forests also offer northern hawk owls an abundance of food in the form of small rodents, which thrive in post-fire habitats, as well as the kind of open foraging habitat these owls prefer and lots of snags for hunting perches. For these owls, fire is not a destructive enemy but an essential part of their environment.

Appearance

This medium-sized owl (length 14–18 inches / 36–45 cm) is distinguished by its hawk-like characteristics, including its long tail and tendency to perch at an angle (most other owls usually sit erect). The plumage is largely dark brown and white: the underparts and tail are barred and the mostly dark brown dorsal plumage is heavily streaked and spotted

■ Year-round
╷╷╷ Southern limits of wintering range

white. The yellow bill and relatively small yellow eyes are set in a small grayish facial disk framed by a broad black crescent on either side; a whitish streak topped by a black spot runs down the hind-neck behind each of the crescents. The crown and nape are variably spotted or mottled black and white, and there is usually a solid black patch on the central nape. The densely feathered legs and feet are light brown.

Voice

Northern hawk owls vocalize frequently during the breeding season, starting as early as February. The male's display call—given in flight or from a prominent perch—is a rapid burbling *ulululul,* continuing for up to 14 seconds and repeated every few seconds or minutes. The female's advertising call is similar, but hoarser, shriller and usually shorter. Various screeching and sharp trilling calls are given to signal alarm or when attacking intruders; they are also used as contact calls and in other social contexts. Pairs perform loud duets before copulating.

Activity Timing and Roost Sites

The northern hawk owl is one of the world's most diurnal owls, typically hunting during daylight hours and rarely, if ever, in complete darkness. During the day these owls often rest on exposed treetop perches. At night they usually sleep on tree limbs, perched close to the trunk and hidden by foliage.

Distribution

This widespread owl's circumpolar range encompasses the boreal forest region of North America and Eurasia. In North America northern hawk owls are sparsely distributed throughout most of Alaska and Canada south of the treeline, except for southern parts of the western provinces. They periodically wander south of the usual breeding range in winter and may stay to nest as far south as Nova Scotia and the extreme northern United States.

Habitat

Northern hawk owls inhabit coniferous and mixed coniferous/deciduous forests, favoring forest edges and open areas such as wetlands, meadows and old burns. They avoid dense stands. Habitat selection is strongly influenced by the abundance of prey. Fire is important for providing habitat for prey, open areas for hunting and suitable nest trees. During irruptions, northern hawk owls sometimes visit woodland and prairie habitats.

Feeding

Voles dominate the diet, especially during breeding season. Other prey include mammals, ranging in size from shrews to hares, and birds, including grouse, ptarmigan and songbirds. Northern hawk owls usually hunt from high perches, swooping down fast and low to strike their victims. They sometimes scavenge dead animals.

Breeding

In North America northern hawk owls are apparently monogamous; rare instances of polygyny have been recorded in Europe. They do not breed in years of prey scarcity. The typical nest site is a cavity in a decayed or fire-hollowed tree, a depression in the top of a broken tree trunk or an old pileated woodpecker or northern flicker cavity. Stick nests made by crows or ravens are occasionally used. The clutch size ranges from three to nine, varying with prey availability. Depending on location, the eggs are laid between late March and late June. They are incubated for 25 to 30 days and the young leave the nest 3 to 5 weeks later. At about 10 to 12 weeks old, juveniles become independent and may begin to disperse.

Migration and Other Movements

Northern hawk owl are nonmigratory but highly nomadic, moving from one breeding area to another as local food supply varies in response to habitat changes (such as post-fire forest regrowth) and prey population cycles. Periodic southward irruptions in winter seem to be linked to snowshoe hare and vole population cycles. Adult females and juveniles disperse earlier and travel farther than adult males.

Conservation

The North American population is probably stable, but numbers and trends are unclear because of this species' largely inaccessible breeding range, low breeding densities, erratic distribution and periodic irruptions. Habitat loss related to forestry and fire suppression is the main threat to the species. Individual northern hawk owls are vulnerable to being shot because they show little fear of humans and perch conspicuously, and to getting caught in leghold traps when they try to retrieve dead prey used as bait.

Acknowledgments

I would like to thank all of the biologists, researchers and naturalists who responded to my requests for information through the CAVNET and Association of Professional Biologists of B.C. list-servs, as well as those who assisted me when I contacted them directly. I particularly appreciated the contributions of John Arvin, Jakob Dulisse, Jared Hobbs, Louis Imbeau, Greg Lasley, Ken Otter, Tania Tripp and Karen Wiebe. Special thanks to Paul Levesque for an enjoyable night spent banding northern saw-whet owls at the Rocky Point Bird Observatory.

Rebecca Raworth of the University of Victoria provided me with invaluable library assistance. Carl Downing, Chuck Otte and Richard Schneider helpfully sent me copies of hard-to-obtain publications.

For their roles in seeing this book through from concept to publication, I thank my literary agent, Carolyn Swayze, and the editorial staff at Firefly Books, especially Barbara Campbell.

And, last but not least, I'm grateful for all the support and encouragement Mark Zuehlke gave me along the way.

Glossary

allopreening – preening of one bird by another; also known as mutual preening. Among owls, allopreening usually involves mates, but sometimes occurs between siblings or parents and young. See also *preening.*

barbs – small rods or filaments that lie parallel to each other and extend outward from a feather shaft, forming the feather vane

brood – all of the young hatched from one clutch of eggs

brooding – a form of parental care. The adult sits over its young to shelter them from the elements and guard them against predators.

brood patch – an area on the belly of a bird that becomes temporarily featherless shortly before the incubation period begins. The skin in the area of the patch also becomes soft and engorged with blood vessels to assist in heat transfer to the eggs and, later, to the nestlings. Sometimes referred to as an incubation patch.

clutch – the full set of eggs laid by one female during a single nesting session

color morph – one of various distinct plumage color variations found within a single species

dorsal – relating to the back of the body

extirpate – to eliminate a species from one part of its range, making it extinct at the local or regional level

fledgling – a young bird that has recently left its nest

foraging – searching for and collecting food

genus (*plural* **genera**) – a taxonomic category consisting of a group of closely related species. The genus provides the first part of an organism's scientific name.

hybridize – to crossbreed with a different species

irruption – a substantial, rapid and temporary increase in population size or density in response to an episodic environmental change that is favorable to that species

mandible – One half, either the upper or lower, of a bird's beak. A mandible consists of a hard, thick sheath of modified skin cells molded around a bony core that is part of the skull.

mobbing – a demonstration by birds of a weaker species against a bird that is perceived as a potential threat or predator

montane – associated with mountainous regions

morph – see *color morph*

nape – the back of the neck

natal down – the first plumage worn by a nestling

natal territory – the place where a bird was hatched

polyandry – a mating strategy in which one female breeds with two or more males in the same time period. Serial polyandry occurs when a female leaves her offspring late in the nestling period to mate with a second male, while her first mate finishes raising the first brood.

polygamy – a mating strategy in which one individual breeds with two or more mates during a single breeding season; includes polyandry and polygyny.

polygyny – a mating strategy in which one male breeds with two or more females in the same time period

preening – a body-maintenance activity in which the bird uses its bill to remove dirt and parasites from its skin and feathers, straighten feather barbs and distribute the oily secretions produced by the uropygial gland through its plumage. See also *allopreening*.

primaries – the flight feathers on the outermost portion of the wing

remiges – the flight feathers of the wing; includes primaries and secondaries.

rectrices – the long flight feathers of the tail

riparian habitat – a distinctive habitat adjacent to a stream, river, lake or wetland. Elevated moisture levels typically make riparian areas lusher and often more treed than neighboring habitats farther from the water.

rufous – reddish

serial polyandry – see *polyandry*

snag – a standing dead tree

tarsus – the lower leg of a bird; the segment to which the toes are attached.

understory – vegetation growing below the main tree canopy in a forest or woodland; includes young trees, shrubs, herbaceous plants and grasses.

uropygial gland – a gland located on the rump just above the base of the tail feathers; also known as the preen gland or oil gland because it produces oils used for preening.

vane – the flat part of a feather; made up of a series of barbs that extend out from the central shaft.

Bibliography

Arsenault, D.P., P.B. Stacey and G.A. Hoelzer. "No Extra-pair Fertilization in Flammulated Owls Despite Aggregated Nesting." *Condor* 104 (2002): 197–201.

Arvin, J. "John Arvin's Comments on Stygian Owl from a Letter to the Discoverers." http://texasbirds.org/tbrc/stowrang.htm.

Audubon, J.J. *The Birds of America*, vol. 1. 1840–44. Reprint, New York: Dover, 1967.

Austing, G.R., and J.B. Holt Jr. *The World of the Great Horned Owl.* Philadelphia: J.B. Lippincott, 1966.

Balgooyen, T.G. "Pygmy Owl Attacks California Quail." *Auk* 86 (1969): 358.

Beaucher, M., and J.A. Dulisse. "First Confirmed Breeding Record for the Western Screech-owl (*Megascops kennicottii macfarlanei*) in Southeastern British Columbia." *Northwestern Naturalist* 85 (2004): 128–30.

Bent, A.C. Life *Histories of North American Birds of Prey, Part Two.* Bulletin 170. Washington, DC: U.S. National Museum, 1938. Reprint, New York: Dover, 1961.

Berger, C. *Owls.* Mechanicsburg, PA: Stackpole Books, 2005.

Bêty, J., G. Gauthier, J-F. Giroux and E. Korpimäki. "Are Goose Nesting Success and Lemming Cycles Linked? Interplay Between Nest Density and Predators." *Oikos* 93 (2001): 388–400.

Bondrup-Nielsen, S. "Ambivalence of the Concealing Pose of Owls." *Canadian Field-Naturalist* 97 (1983): 329–31.

———. "Thawing of Frozen Prey by Boreal and Saw-whet Owls." *Canadian Journal of Zoology* 55 (1977): 595–601.

Brandt, H. "Some Arizona Bird Studies." *Auk* 54 (1937): 62–64.

Breen, T.F., and J.W. Parrish Jr. "Eastern Screech-owl Hatches an American Kestrel." *Journal of Field Ornithology* 67 (1996): 612–13.

Brinker, D.F., K.E. Duffy, D.M. Whalen, B.D. Watts and K.M. Dodge. "Autumn Migration of Northern Saw-whet Owls (*Aegolius acadicus*) in the Middle Atlantic and Northeastern United States: What Observations from 1995 Suggest." In *Biology and Conservation of Owls of the Northern Hemisphere: Second International Symposium,* 74–86. USDA Forest Service, General Technical Report NC-190, 1997.

Bull, E., and M.G. Henjum. *Ecology of the Great Gray Owl.* USDA Forest Service, General Technical Report PNW-265, 1990.

Bunn, D.S., A.B. Warburton and R.D.S. Wilson. *The Barn Owl.* Staffordshire, UK: T. & A.D. Poyser, 1982.

Cade, T.J. "A Hawk Owl Bathing with Snow." *Condor* 54 (1952): 360.

Campbell, B. "Bird Notes from Southern Arizona." *Condor* 36 (1934): 201–3.

Center for Biological Diversity and Defenders of Wildlife. *Petition to List the Cactus Ferruginous Pygmy Owl as a Threatened or Endangered Species under the Endangered Species Act.* March 15, 2007.

Cheveau, M., P. Drapeau, L. Imbeau and Y. Bergeron. "Owl Winter Irruptions as an Indicator of Small Mammal Population Cycles in the Boreal Forest of Eastern North America." *Oikos* 107 (2004): 190–98.

Dark, S.J., R.J. Gutiérrez and G.I. Gould Jr. "The Barred Owl (*Strix varia*) Invasion in California." *Auk* 115 (1998): 50–56.

Davies, J.M., and M. Restani. "Survival and Movements of Juvenile Burrowing Owls During the Postfledging Period." *Condor* 108 (2006): 282–91.

del Hoyo, J., A. Elliott and J. Sargatal, eds. *Handbook of the Birds of the World.* Vol. 5, *Barn-owls to Hummingbirds.* Barcelona: Lynx Edicions, BirdLife International, 2002.

Deppe, C., D. Holt, J. Tewksbury, L. Broberg, J. Petersen and K. Wood. "Effect of Northern Pygmy-owl (*Glaucidium gnoma*) Eyespots on Avian Mobbing." *Auk* 120 (2003): 765–71.

Dice, L.R. "Minimum Intensities of Illumination under Which Owls Can Find Dead Prey by Sight." *American Naturalist* 79 (1945): 385–416.

Duncan, J.R., D.H. Johnson and T.H. Nicholls, eds. *Biology and Conservation of Owls of the Northern Hemisphere: Second International Symposium.* USDA Forest Service, General Technical Report NC-190, 1997.

Ehrlich, P.R., D.S. Dobkin and D. Wheye. *The Birder's Handbook.* New York: Simon and Schuster, 1988.

Enríquez-Rocha, P., J.L. Rangel-Salazar and D.W. Holt. "Presence and Distribution of Mexican Owls: A Review." *Journal of Raptor Research* 27 (1993): 154–60.

Flesch, A.D., and R.J. Steidl. "Population Trends and Implications for Monitoring Cactus Ferruginous Pygmy Owls in Northern Mexico." *Journal of Wildlife Management* 70 (2006): 867–71.

Forsman, E.D. "Lessons Learned in 30 Years of Research and Management on the Northern Spotted Owl (*Strix occidentalis*)." In *Ecology and Conservation of Owls,* 277–85. Collingwood, Australia: CSIRO Publishing, 2002.

———. "Natal and Post-natal Dispersal of Northern Spotted Owls (*Strix occidentalis*)." In *Ecology and Conservation of Owls,* 56–57. Collingwood, Australia: CSIRO Publishing, 2002.

Forsman, E.D., A. Giese, D. Manson and S. Sovern. "Renesting by Spotted Owls." *Condor* 97 (1995): 1078–80.

Forsman, E.D., and H.M. Wight. "Allopreening in Owls: What Are Its Functions?" *Auk* 96 (1979): 525–31.

Frost, B.J., P.J. Baldwin and M. Csizy. "Auditory Localization in the Northern Saw-whet Owl, *Aegolius acadicus*." *Canadian Journal of Zoology* 67 (1989): 1955–59.

Frye, G.G., and R.P. Gerhardt. "Northern Saw-whet Owl (*Aegolius acadicus*) Migration in the Pacific Northwest." *Western North American Naturalist* 63 (2003): 353–57.

Gehlbach, F.R., and R.S. Baldridge. "Live Blind Snakes (*Leptotyphlops dulcis*) in Eastern Screech Owl (*Otus asio*) Nests: A Novel Commensalism." *Oecologia* 71 (1987): 560–63.

Gerhardt, R.P. "Response of Mottled Owls to Broadcast of Conspecific Call." *Journal of Field Ornithology* 62 (1991): 239–44.

Gerhardt, R.P., D.M. Gerhardt, C.J. Flatten and N.B. González. "The Food Habits of Sympatric *Ciccaba* Owls in Northern Guatemala." *Journal of Field Ornithology* 65 (1994): 258–64.

Gerhardt, R.P., N.B. González, D.M. Gerhardt and C.J. Flatten. "Breeding Biology and Home Range of Two *Ciccaba* Owls." *Wilson Bulletin* 106 (1994): 629–39.

Gervais, J.A., and D.K. Rosenberg. "Western Burrowing Owls in California Produce Second Broods of Chicks." *Wilson Bulletin* 111 (1999): 569–71.

Gessaman, J.A. "Bioenergetics of the Snowy Owl (*Nyctea scandiaca*)." *Arctic and Alpine Research* 4 (1972): 223–38.

Grimm, R.J., and W.M. Whitehouse. "Pellet Formation in a Great Horned Owl: A Roentgenographic Study." *Auk* 80 (1963): 301–6.

Gutiérrez, R.J., and G.F. Barrowclough. "Redefining the Distributional Boundaries of the Northern and California Spotted Owls: Implications for Conservation." *Condor* 107 (2005): 182–87.

Hannah, K.C, and J.S. Hoyt. "Northern Hawk Owls and Recent Burns: Does Burn Age Matter?" *Condor* 106 (2004): 420–23.

Hayward, G.D., and J. Verner, eds. *Flammulated, Boreal, and Great Gray Owls in the United States: A Technical Conservation Assessment.* USDA Forest Service, General Technical Report RM-253, 1994.

Hayward, J.L., J.G. Galusha and G. Frias. "Analysis of Great Horned Owl Pellets with Rhinoceros Auklet Remains." *Auk* 110 (1993): 133–35.

Highfill, K. "Great Horned Owls (*Bubo virginianus*) Successful Nesting in a Flower Pot in Northern Kansas." *Kansas Ornithological Society Bulletin* 48 (1997): 23–24.

Hobson, K.A., and S.G. Sealy. "Marine Protein Contributions to the Diet of Northern Saw-whet Owls on the Queen Charlotte Islands: A Stable-Isotope Approach." *Auk* 108 (1991): 437–40.

Hocking, B., and B.L. Mitchell. "Owl Vision." *Ibis* 103 (1961): 284–88.

Hoffman, W., G.E. Woolfenden and P.W. Smith. "Antillean Short-eared Owls Invade Southern Florida." *Wilson Bulletin* 111 (1999): 303–13.

Houston, C.S. "Barred Owl Nest in Attic of Shed." *Wilson Bulletin* 111 (1999): 272–73.

———. "Observation and Salvage of Great Horned Owl Nest." *Blue Jay* 23 (1965): 164–65.

Howell, A.B. "Some Results of a Winter's Observations in Arizona." *Condor* 18 (1916): 209–14.

Howell, S.N.G., and M.B. Robbins. "Species Limits of the Least Pygmy-owl (*Glaucidium minutissimum*) Complex." *Wilson Bulletin* 107 (1995): 7–25.

Howell, S.N.G., and S. Webb. *A Guide to the Birds of Mexico and Northern Central America.* Oxford: Oxford University Press, 1995.

Johnson, R.R., J-L.E. Cartron, L.T. Haight, R.B. Duncan and K.J. Kingsley. "Cactus Ferruginous Pygmy-owl in Arizona, 1872–1971." *Southwestern Naturalist* 48 (2003): 389–401.

Kelly, E.G., and E.D. Forsman. "Recent Records of Hybridization Between Barred Owls (*Strix varia*) and Northern Spotted Owls (*S. occidentalis caurina*)." *Auk* 121 (2004): 806–10.

Kelly, E.G., E.D. Forsman and R.G. Anthony. "Are Barred Owls Displacing Spotted Owls?" *Condor* 105 (2003): 45–53.

Kelso, L., and E.H. Kelso. "The Relation of Feathering of Feet of American Owls to Humidity of Environment and to Life Zones." *Auk* 53 (1936): 51–56.

Klute, D.S., et al. *Status Assessment and Conservation Plan for the Western Burrowing Owl in the United States.* Washington: USDI Fish and Wildlife Service, Biological Technical Publication BTP-R6001-2003, 2003.

Konishi, M. "Listening with Two Ears." *Scientific American* 268 (1993): 66–73.

Korfanta, N.M., D.B. McDonald and T.C. Glenn. "Burrowing Owl (*Athene cunicularia*) Population Genetics: A Comparison of North American Forms and Migratory Habits." *Auk* 122 (2005): 464–78.

Kotler, B.P. "Owl Predation on Desert Rodents Which Differ in Morphology and Behavior." *Journal of Mammalogy* 66 (1985): 824–28.

Lasley, G.W., C. Sexton and D. Hillsman. "First Record of Mottled Owl (*Ciccaba virgata*) in the United States." *American Birds* 42 (1988): 23–24.

Lawless, S.G., G. Ritchison, P.H. Klatt and D.F. Westneat. "The Mating Strategies of Eastern Screech-owls: A Genetic Analysis." *Condor* 99 (1997): 213–17.

Lignon, J.D. "Some Aspects of Temperature Relations in Small Owls." *Auk* 86 (1969): 458–72.

Linkhart, B.D., and R.T. Reynolds. "Longevity of Flammulated Owls: Additional Records and Comparisons to Other North American Strigiforms." *Journal of Field Ornithology* 75 (2004): 192–95.

Maples, M.T., D.W. Holt and R.W. Campbell. "Ground-nesting Long-eared Owls." *Wilson Bulletin* 107 (1995): 563–65.

Marks, J.S., J.L. Dickinson and J. Haydock. "Genetic Monogamy in Long-eared Owls." *Condor* 101 (1999): 854–59.

———. "Serial Polyandry and Alloparenting in Long-eared Owls." *Condor* 104 (2002): 202–4.

Marks, J.S., J.H. Doremus and R.J. Cannings. "Polygyny in the Northern Saw-whet Owl." *Auk* 106 (1989): 732–34.

Marshall, J.T. "Birds of the Pine-Oak Woodland in Southern Arizona and Adjacent Mexico." *Pacific Coast Avifauna* 32 (1957).

Marti, C.D. "Feeding Ecology of Four Sympatric Owls." *Condor* 76 (1974): 45–61.

———. "Same-Nest Polygyny in the Barn Owl." *Condor* 92 (1990): 261–63.

Martin, D.J. "Selected Aspects of Burrowing Owl Ecology and Behavior." *Condor* 75 (1973): 446–56.

———. "A Spectrographic Analysis of Burrowing Owl Vocalizations." *Auk* 90 (1973): 564–78.

Martin, G. *Birds by Night.* London: T. & A.D. Poyser, 1990.

Martin, G.R. "An Owl's Eye: Schematic Optics and Visual Performance in *Strix aluco.*" *Journal of Comparative Physiology* A. 145 (1982): 341–49.

———. "Sensory Capacities and the Nocturnal Habit of Owls (*Strigiformes*)." *Ibis* 128 (1986): 266–77.

McCafferty, D.J., J.B. Montcrieff and I.R. Taylor. "How Much Energy Do Barn Owls (*Tyto alba*) Save by Roosting?" *Journal of Thermal Biology* 26 (2001): 193–203.

McQueen, L.B. "Observations on Copulatory Behavior of a Pair of Screech Owls (*Otus asio*)." *Condor* 74 (1972): 101.

Miller, A.H. "The Vocal Apparatus of Some North American Owls." *Condor* 36 (1934): 204–13.

———. "The Vocal Apparatus of Two South American Owls." *Condor* 65 (1963): 440–41.

———. "The Structural Basis of the Voice of the Flammulated Owl." *Auk* 64 (1947): 133–35.

Millsap, B.A., and C. Bear. "Double-Brooding by Florida Burrowing Owls." *Wilson Bulletin* 102 (1990): 313–17.

Murphy, C.J., and H.C. Howland. "Owl Eyes: Accommodation, Corneal Curvature and Refractive State." *Journal of Comparative Physiology* 151 (1983): 277–84.

Murray, G.A. "Geographic Variation in the Clutch Sizes of Seven Owl Species." *Auk* 93 (1976): 602–13.

Nelson, E.W. "Descriptions of Five New Birds from Mexico." *Auk* 18 (1901): 46–49.

Nero, R.W. *The Great Gray Owl: Phantom of the Northern Forest.* Washington, DC: Smithsonian Institution Press, 1980.

Nero, R.W., R.J. Clark, R.J. Knapton and R.H. Hamre, eds. *Biology and Conservation of Northern Forest Owls: Symposium Proceedings.* USDA Forest Service, General Technical Report RM-142, 1987.

Newton, I., R. Kavanagh, J. Olsen and I. Taylor, eds. *Ecology and Conservation of Owls.* Collingwood, Australia: CSIRO Publishing, 2002.

Norberg, R.Å. "Evolution, Structure, and Ecology of Northern Forest Owls." In *Biology and Conservation of Northern Forest Owls: Symposium Proceedings,* 9–43. USDA Forest Service, General Technical Report RM-142, 1987.

———. "Independent Evolution of Outer Ear Asymmetry among Five Owl Lineages; Morphology, Function and Selection. In *Ecology and Conservation of Owls,* 329–42. Collingwood, Australia: CSIRO Publishing, 2002.

Oleyar, M.D, C.D. Marti and M. Mika. "Vertebrate Prey in the Diet of Flammulated Owls in Northern Utah." *Journal of Raptor Research* 37 (2003): 244–46.

Otter, K. "Individual Variation in the Advertising Call of Male Northern Saw-whet Owls." *Journal of Field Ornithology* 67 (1996): 398–405.

Parmalee, D.F. "Canada's Incredible Arctic Owls." *Beaver* (Summer 1972): 30–41.

Perrone, M., Jr. "Adaptive Significance of Ear Tufts in Owls." *Condor* 83 (1981): 383–84.

Perry, R.W., R.E. Brown and D.C. Rudolph. "Mutual Mortality of Great Horned Owl and Southern Black Racer: A Potential Risk of Raptors Preying on Snakes." *Wilson Bulletin* 113 (2001): 345–47.

Poole, A., and F. Gill, eds. *The Birds of North America.* Philadelphia: Academy of Natural Sciences; Washington, DC: American Ornithologists' Union. [This series includes 19 owl accounts.]

Proudfoot, G.A., S.L. Beasom and F. Chavez-Ramirez. *Biology of Ferruginous Pygmy-owls in Texas and Application of Artificial Nest Structures.* Kingsville, TX: Caesar Kleberg Wildlife Research Institute, Texas A&M University, 1999.

Proudfoot, G.A., R.L. Honeycutt and R.D. Slack. "Mitochondrial DNA Variation and Phylogeography of the Ferruginous Pygmy-owl (*Glaucidium brasilianum*)." *Conservation Genetics* 7 (2006): 1–12.

Quigley, R. "Unusual Barn Owl Nest Location." *Wilson Bulletin* 56 (1954): 315.

Rohner, C., J.N.M. Smith, J. Stroman, M. Joyce, F.I. Doyle and R. Boonstra. "Northern Hawk-owls in the Nearctic Boreal Forest: Prey Selection and Population Consequences of Multiple Prey Cycles." *Condor* 97 (1995): 208–20.

Roulin, A., T.W. Jungi, H. Pfister and C. Dijkstra. "Female Barn Owls (*Tyto alba*) Advertise Good Genes." *Proceedings: Biological Sciences* 267 (May 7, 2000): 937–41.

Roulin, A., C. Riols, C. Dijkstra and A-L. Ducrest. "Female Plumage Spottiness Signals Parasite Resistance in the Barn Wwl (*Tyto alba*)." *Behavioral Ecology* 12 (2001): 103–10.

Russell, R.W., P. Dunne, C. Sutton and P. Kerlinger. "A Visual Study of Migrating Owls at Cape May Point, New Jersey." *Condor* 93 (1991): 55–61.

Seidensticker, M.T., D.T. Tyler Flockhart, D.W. Holt and K. Gray. "Growth and Plumage Development of Nestling Long-eared Owls." *Condor* 108 (2006): 981–85.

Smith, D.G., and E. Hiestand. "Alloparenting at an Eastern Screech-owl Nest." *Condor* 92 (1990): 246–47.

Stahlecker, D.W., and R.B. Duncan. "The Boreal Owl at the Southern Terminus of the Rocky Mountains: Undocumented Longtime Resident or Recent Arrival?" *Condor* 98 (1996): 153–61.

Stangl, F.B., M.M. Shipley, J.R. Goetze and C. Jones. "Comments on the Predator–Prey Relationship of the Texas Kangaroo Rat (*Dipodomys elator*) and Barn Owl (*Tyto alba*)." *American Midland Naturalist* 153 (2005): 135–41.

Stock, S.L., P.J. Heglund, G.S. Kaltenecker, J.D. Carlisle and L. Leppert. "Comparative Ecology of the Flammulated Owl and Northern Saw-whet Owl During Fall Migration." *Journal of Raptor Research* 40 (2006): 120–29.

Taverner, P.A., and B.H. Swales. "Notes on the Migration of the Saw-whet Owl." *Auk* 28 (1911): 329–34.

Texas Bird Records Committee. "Stygian Owl." http://texasbirds.org/tbrc/stygowl.htm.

Thomsen, L. "Behavior and Ecology of Burrowing Owls on the Oakland Municipal Airport." *Condor* 73 (1971): 177–92.

Tremblay, J-P., G. Gauthier, D. Lepage and A. Desrochers. "Factors Affecting Nesting Success in Greater Snow Geese: Effects of Habitat and Association with Snowy Owls." *Wilson Bulletin* 109 (1997): 449–61.

Walk, J.W. "Winter Roost Sites of Northern Harriers and Short-eared Owls on Illinois Grasslands." *Journal of Raptor Research* 32 (1998): 116–19.

Walk, J.W., T.L. Esker and S.A. Simpson. "Continuous Nesting of Barn Owls in Illinois." *Wilson Bulletin* 111 (1999): 572–73.

Watson, A. "The Behaviour, Breeding, and Food-Ecology of the Snowy Owl *Nyctea scandianca*." *Ibis* 99 (1957): 419–62.

Welty, J.C. *The Life of Birds*. Philadelphia: Saunders College Publishing, 1979.

Whalen, D.M., and B.D. Watts. "Annual Migration Density and Stopover Patterns of Northern Saw-whet Owls (*Aegolius acadicus*)." *Auk* 119 (2002): 1154–61.

Whitfield, C.J. "A Screech Owl Captured by a Snake." *Condor* 36 (1934): 84.

Wiebe, K.L. "Northernmost Nest Record of the Flammulated Owl (*Otus flammeolus*) in Canada." *Wildlife Afield* (2005): 78–79.

Zambrano, R. "The First Record of Burrowing Owls Nesting in a Building." *Wilson Bulletin* 110 (1998): 560–61.

Literary Permissions

The publisher is grateful to the following authors and publishers for permission to reproduce excerpts from their work. Every effort has been made to determine the sources and give proper credit. Any oversight or error is unintentional.

Page 38: "When I picked them up…" from "Pygmy Owl Attacks California Quail," *Auk* 86 (1969), by Thomas Balgooyen. Reproduced with kind permission of the American Ornithologists' Union.

Page 40: "Through binoculars one can see…" from *The Barn Owl,* by D.S. Bunn, A.B. Warburton and R.D.S. Wilson (1982). Reproduced with permission of T&AD Poyser, an imprint of A&C Black.

Page 43: "Typically, an owl would fly…" from *The Great Gray Owl: Phantom of the Northern Forest,* by Robert W. Nero (1980). Reproduced with kind permission of the author.

Page 45: "the owl picks [the prey] up…" from *The Barn Owl,* by D.S. Bunn, A.B. Warburton and R.D.S. Wilson (1982). Reproduced with permission of T&AD Poyser, an imprint of A&C Black.

Page 59: "to hear these rending screams…" from "Canada's Incredible Arctic Owls," *The Beaver,* vol. 52, no. 1 (summer 1972), by David Parmalee. Reproduced with kind permission of *The Beaver.*

Page 67: "Like some miserable, tortured clown…" from "Canada's Incredible Arctic Owls," *The Beaver,* vol. 52, no. 1 (summer 1972), by David Parmalee. Reproduced with kind permission of *The Beaver.*

Page 111: "Consequently, there is a great din…" from "Birds of the Pine-Oak Woodland in Southern Arizona and Adjacent Mexico," *Pacific Coast Avifauna,* vol. 32 (1957), by Joe T. Marshall Jr. Reproduced with kind permission of the Cooper Ornithological Society. All rights reserved.

Page 116: "…commenced a vigorous bathing…" from "A Hawk Owl Bathing with Snow," *Condor,* vol. 54 (1952), by Tom J. Cade. Reproduced with kind permission of the Cooper Ornithological Society. All rights reserved.

Photo Credits

Front cover © Tom Vezo/Minden Pictures
Back cover © Rolf Kopfle/ardea.com
Title page © Paul Nicklen/National Geographic/ Getty Images

6 © Jeff Haynes/AFP/Getty Images
10 © Jared Hobbs
13 © John Lowman/Viewpoints West
14 © W. Perry Conway/CORBIS
17 © Jared Hobbs
18 © John Lowman/Viewpoints West
20 © W. Perry Conway/CORBIS
22 © W. Perry Conway/CORBIS
25 © Yva Momatiuk/John Eastcott/Minden Pictures
27 © Joe McDonald/CORBIS
31 © Wilderness Light
33 © W. Perry Conway/CORBIS
34 © Joe McDonald/CORBIS
37 © Tom & Pat Leeson/ardea.com
38 © John Lowman/Viewpoints West
41 © Tim Zurowski/CORBIS
42 © Rick & Nora Bowers/VIREO
44 © Rolf Kopfle/ardea.com
46 © NHPA/Karl Switak
49 © Roberta Olenick/Viewpoints West
50 © G. Pecker/Ivy Images
52 © Joe McDonald/CORBIS
57 © Kennan Ward/CORBIS
58 © Roberta Olenick/Viewpoints West
62 © Michael Quinton/Minden Pictures
65 © Fritz Poelking/DRK PHOTO
66 © W. Perry Conway/CORBIS
68 © Michael Quinton/Minden Pictures
72 © Konrad Wothe/Minden Pictures
76 © Michael Quinton/Minden Pictures
79 © G. Pecker/Ivy Images
81 © DRK PHOTO
82 © Michio Hoshino/Minden Pictures
85 © Steve Kaufman/DRK PHOTO
86 © Michael Quinton/Minden Pictures
88 © G. Pecker/Ivy Images
90 © Michael Quinton/Minden Pictures
93 © G. Pecker/Ivy Images

94 © W. Perry Conway/CORBIS
97 © Rolf Nussbaumer/npl/Minden Pictures
99 © Jukka Jantunen/VIREO
100 © Shattil & Rozinski/naturepl.com
103 © Wayne Lynch/DRK PHOTO
106 © Vincent Munier/naturepl.com
109 © Thomas Mangelsen/Minden Pictures
110 © Tim Zurowski/CORBIS
112 © Greg Lasley
114 © Donald M. Jones/DRK PHOTO
117 © Michael Quinton/Minden Pictures
118 © Joe McDonald/CORBIS
123 © Rick & Nora Bowers/Bowers Photo
125 © W. Perry Conway/CORBIS
128 © Barrett & MacKay Photography Inc.
130 © Tim Fitzharris/Minden Pictures
135 © Gerard Bailey/VIREO
138 © Bill Ivy/Ivy Images
143 © Rick & Nora Bowers/Bowers Photo
146 © Roberta Olenick/Viewpoints West
151 © W. Perry Conway/CORBIS
152 © Tim Fitzharris/Minden Pictures
157 © Adrian Binns/VIREO
159 © Petr Myska/imanat.com
162 © Rolf Nussbaumer/npl/Minden Pictures
165 © Alan and Sandy Carey/Ivy Images
169 © David Welling/naturepl.com
170 © Rick & Nora Bowers/Bowers Photo
173 © Tim Fitzharris/Minden Pictures
177 © Jared Hobbs
180 © Rick & Nora Bowers/Bowers Photo
183 © Rick & Nora Bowers/Bowers Photo
186 © Greg Lasley
191 © D. Robert & Lorri Franz/CORBIS
193 © Jared Hobbs
197 © NHPA/Thomas Kitchin & Victoria Hurst

Illustrations: Imagineering Media Services Inc.

Maps reprinted with permission from the Cornell Lab of Ornithology's *Birds of North America Online*, www.birds.cornell.edu/bna

Index

Species profiles are indicated in **bold** type.